Dimensions Math®
Textbook 2A

Authors and Reviewers

Bill Jackson

Jenny Kempe

Cassandra Turner

Allison Coates

Tricia Salerno

Pearly Yuen

Consultant

Dr. Richard Askey

Singapore Math Inc.

Published by Singapore Math Inc.

19535 SW 129th Avenue

Tualatin, OR 97062

www.singaporemath.com

Dimensions Math® Textbook 2A

ISBN 978-1-947226-06-7

First published 2018

Reprinted 2019, 2020

Printed in China

Acknowledgments

Editing by the Singapore Math Inc. team.

Design and illustration by Cameron Wray with Carli Fronius.

Preface

The Dimensions Math® Pre-Kindergarten to Grade 5 series is based on the pedagogy and methodology of math education in Singapore. The curriculum develops concepts in increasing levels of abstraction, emphasizing the three pedagogical stages: Concrete, Pictorial, and Abstract. Each topic is introduced, then thoughtfully developed through the use of problem solving, student discourse, and opportunities for mastery of skills.

Features and Lesson Components

Students work through the lessons with the help of five friends: Emma, Alex, Sofia, Dion, and Mei. The characters appear throughout the series and help students develop metacognitive reasoning through questions, hints, and ideas.

The colored boxes and blank lines in the textbook lessons are used to facilitate student discussion. Rather than writing in the textbooks, students can use whiteboards or notebooks to record their ideas, methods, and solutions.

Chapter Opener

Each chapter begins with an engaging scenario that stimulates student curiosity in new concepts. This scenario also provides teachers an opportunity to review skills.

Think

Students, with guidance from teachers, solve a problem using a variety of methods.

Learn

One or more solutions to the problem in **Think** are presented, along with definitions and other information to consolidate the concepts introduced in **Think**.

Do

A variety of practice problems allow teachers to lead discussion or encourage independent mastery. These activities solidify and deepen student understanding of the concepts.

Exercise

A pencil icon  at the end of the lesson links to additional practice problems in the workbook.

Practice

Periodic practice provides teachers with opportunities for consolidation, remediation, and assessment.

Review

Cumulative reviews provide ongoing practice of concepts and skills.

Emma   Alex   Sofia   Dion   Mei

Contents

Chapter	Lesson	Page

Chapter	Lesson	Page

Chapter	Lesson	Page

Chapter 1

Numbers to 1,000

Think

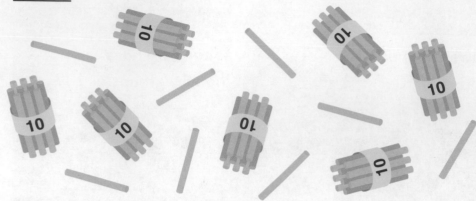

How many straws are there?

Learn

7 tens and 8 ones make 78.

70 and 8 make ____.

8 more than 70 is ____.

70 + 8 = ____

There are ____ straws.

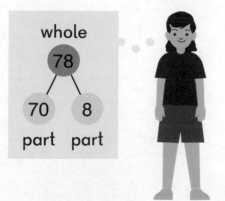

seventy-eight

Do

1 (a) How many bananas are there?

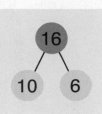

1 ten and 6 ones make 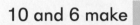 .

10 and 6 make 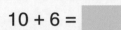 .

10 + 6 =

There are bananas.

(b) How many pens are there?

5 tens and ones make 53.

50 and make 53.

50 + = 53

There are pens.

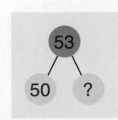

(c) How many 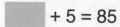 are there?

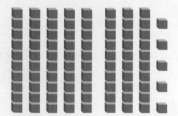

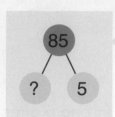

[   ] tens and 5 ones make 85.

[   ] and 5 make 85.

[   ] + 5 = 85

There are [   ] .

(d) How many straws are there?

ten, twenty, thirty, forty, ... , one hundred

[   ] tens make 100.

There are [   ] straws.

2 Show the numbers.
How many tens and ones are in each number?

(a) 19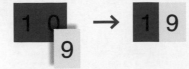

(b) 32

(c) 23

(d) 57

(e) 75

(f) 88

(g) 90

3 (a) 80 + 4 = ▢

(b) 4 + 80 = ▢

(c) 84 − 4 = ▢

(d) 84 − 80 = ▢

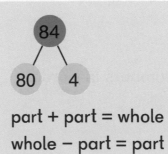

part + part = whole
whole − part = part

4 (a) ▢ + 3 = 63

(b) ▢ + 70 = 72

(c) 39 − ▢ = 30

(d) ▢ − 50 = 5

5 Write the numbers.

(a) twenty-five

(b) fifty-two

(c) nineteen

(d) ninety-one

(e) eighty

(f) 10 tens

(g) 6 tens and 2 ones

(h) 6 ones and 2 tens

6 Write the numbers in words.

(a) 14

(b) 41

(c) 95

(d) 58

(e) 33

(f) 100

Think

What number is 1 more than 48?
What number is 3 more than 48?
What number is 2 less than 48?
What number is 30 more than 48?
What number is 20 less than 48?

Learn

1 more than 48 is 49.

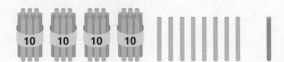

48 + 1 = ▭

Count on by ones from **48**: 49.

3 more than 48 is 51.

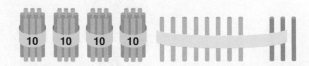

48 + 3 = ▭

Count on by ones from **48**:
49, 50, 51.

2 less than 48 is 46.

48 − 2 = ▢

Count back by ones from **48**:
47, 46.

---

30 more than 48 is 78.

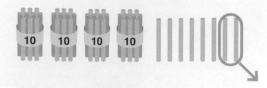

48 + 30 = ▢

Count on by tens from **48**:
58, 68, 78.

---

20 less than 48 is 28.

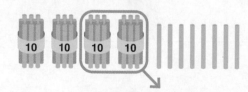

48 − 20 = ▢

Count back by tens from **48**:
38, 28.

Do

1 What number is ...

(a) 1 more than 59

(b) 1 less than 59

(c) 10 more than 59

(d) 10 less than 59

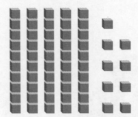

2 What number is ...

(a) 2 more than 71

(b) 2 less than 71

(c) 20 more than 71

(d) 20 less than 71

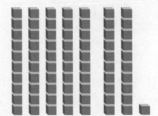

3 What number is ...

(a) 3 more than 68

(b) 3 less than 68

(c) 30 more than 68

(d) 30 less than 68

4 (a) 35 + 3 = ☐   35 + 30 = ☐

(b) 87 − 2 = ☐   87 − 20 = ☐

(c) 49 + 1 = ☐   49 + 10 = ☐

(d) 60 − 2 = ☐   60 − 20 = ☐

(e) 82 − 3 = ☐   82 − 30 = ☐

Exercise 2 • page 3

Think

How can we compare the number of artichokes in the boxes?

Learn

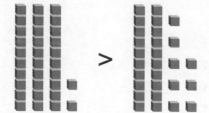

3 tens is greater than 2 tens, so ...

32 is greater than 28.

32 > 28

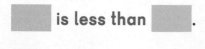 is less than .

⬜ < ⬜

The signs > (is greater than) and < (is less than) show the relationship between the values of two unequal numbers.

Do

1 What sign, > or <, goes in the ◯?

(a)

Tens	Ones
4	9

Tens	Ones
6	1

4 tens is less than 6 tens.

49 ◯ 61

(b)

Tens	Ones

Tens	Ones

The tens are the same, so ...

58 ◯ 53

2 Compare the number of straws.

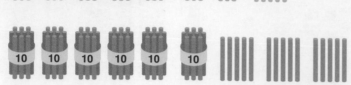

When the amounts are the same,
we use the equal sign, **=**.

3 What sign, >, <, or =, goes in the ◯?

(a) 85 ◯ 34

(b) 47 ◯ 67

(c) 69 ◯ 96

(d) 45 ◯ 5

(e) eight ◯ eighteen

(f) sixty-seven ◯ seventy-four

(g) 5 ones 4 tens ◯ 1 ten 4 ones

(h) 5 tens 18 ones ◯ 6 tens 8 ones

4 Put the numbers in order from least to greatest.

(a)

| 37 | 57 | 27 |

(b)

| 74 | 72 | 79 | 70 |

(c)

| 55 | 45 | 54 | 44 | 40 |

(d)

| 32 | 57 | 7 | 23 | 38 | 83 |

5 What sign, >, <, or =, goes in the ◯?

(a) 80 + 7 ◯ 7 + 80

(b) 3 + 20 ◯ 4 + 10

(c) 32 − 30 ◯ 32 − 2

(d) 69 + 3 ◯ 29 + 30

(e) 47 − 10 ◯ 35 + 3

Think

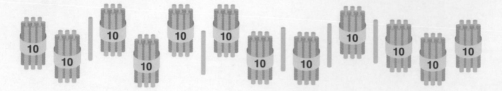

There are more than 10 tens.

How many straws are there?

Learn

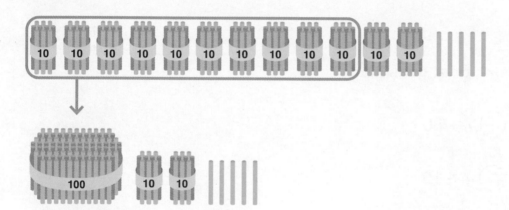

100, 110, 120, 121, 122, ...

one hundred twenty-five

10 tens make ▢ hundred.

1 hundred, 2 tens, and 5 ones make 125.

There are ▢ straws.

Do

1 How many straws are there?

(a)

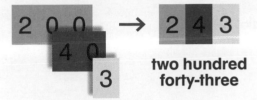

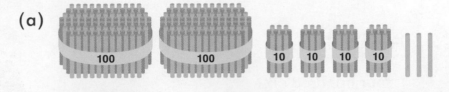

```
2 0 0        2 4 3
  4 0
    3      two hundred
             forty-three
```

100, 200, 210, 220, 230, 240, 241, ...

2 hundreds, 4 tens, and 3 ones make ____.

There are ____ straws.

(b)

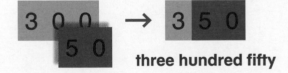

```
3 0 0        3 5 0
  5 0
           three hundred fifty
```

100, 200, 300, 310, ...

3 hundreds and 5 tens make ____.

There are ____ straws.

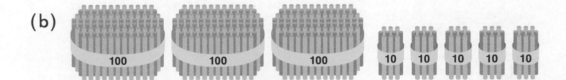

(c)

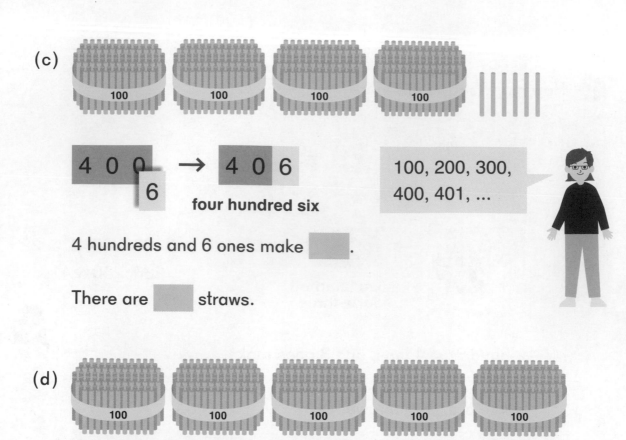

4 0 0 → 4 0 6
  6   **four hundred six**

100, 200, 300, 400, 401, ...

4 hundreds and 6 ones make ▢.

There are ▢ straws.

(d)

1, 0 0 0

one thousand

10 hundreds is ▢.

There are ▢ straws.

2

9 ones, 6 tens, and 2 hundreds is ⬜.

There are ⬜ straws.

3 Show each number with place-value cards.

(a) 358

(b) 630

(c) 809

4 (a) 7 hundreds, 3 tens, and 6 ones make ⬜.

(b) 8 hundreds and 5 tens make ⬜.

(c) 4 ones and 2 hundreds make ⬜.

(d) 3 tens, 9 hundreds, and 1 one make ⬜.

5 (a) 4 hundreds, 4 tens, and 13 ones make ⬜.

(b) 3 hundreds, 12 tens, and 5 ones make ⬜.

Exercise 4 • page 7

Think

How many sticky notes are there?

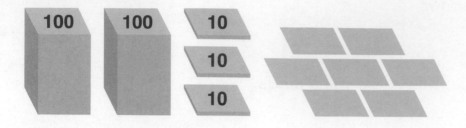

Learn

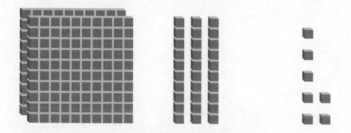

Hundreds	Tens	Ones
2	3	7

two hundred thirty-seven

2 hundreds + 3 tens + 7 ones = 237

$200 + 30 + 7 = $ ▢

There are ▢ sticky notes.

We can also show how many hundreds, tens, and ones this way.

Do

1 (a) 10 ones = [ ] ten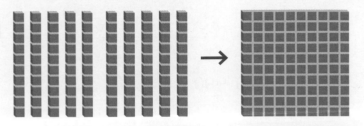

(b) 10 tens = [ ] hundred

(c) 1 ten = [ ] ones

(d) 1 hundred = [ ] tens

(e) 10 tens = [ ] hundred

2 (a)

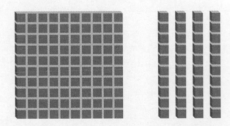

Hundreds	Tens	Ones
1	4	2

100 + 40 + 2 = [ ]

(b)

Hundreds	Tens	Ones

500 + 90 + 6 =

3 (a)

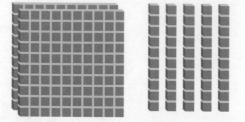

Hundreds	Tens	Ones

200 + 50 =

(b)

Hundreds	Tens	Ones

800 + 80 =

(a)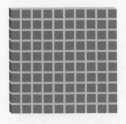

| 100 | | 1 1 1 1 1
1 1 1 1 |

Hundreds	Tens	Ones

1 0 0
9

100 + 9 = ⬜

(b)

| 100 100 100 100 100
100 100 | | 1 1 1 1 1 |

Hundreds	Tens	Ones

7 0 0
5

700 + 5 = ⬜

5

| | 10 10 10 10 10
10 10 10 10 | 1 1 1 1 1
1 |

Hundreds	Tens	Ones

90 + 6 = ⬜

We do not write 096.

6 What number is shown?

(a)
100 100 100 100 100 100 100	10 10 10 10	1 1 1 1 1 1 1 1 1

(b)
100 100 100 100	10 10 10 10 10 10 10 10	

(c)
100 100 100 100 100 100		1 1 1 1 1

7 (a) $500 + 80 + 7 =$ ☐   (b) $600 + 5 =$ ☐

(c) $800 + 80 + 6 =$ ☐   (d) $7 + 300 + 30 =$ ☐

(e) $200 +$ ☐ $= 240$ (f) $200 +$ ☐ $= 204$

(g) $10 + 8 + 400 =$ ☐   (h) $50 + 700 =$ ☐

8 Emma received money for her birthday.
How much did she receive?

She received $ ☐ .

Lesson 6
Comparing Hundreds, Tens, and Ones

Think

$142

Video game console A

$213

Video game console B

Which game console costs less?

Learn

Video game console A

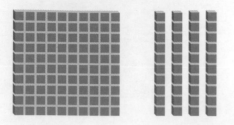

$142

Hundreds	Tens	Ones
1	4	2

Video game console B

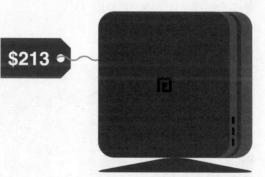

$213

Hundreds	Tens	Ones
2	1	3

142 < 213

Video game console _____ costs less.

1 hundred is less than 2 hundreds.

Do

1 What sign, > or <, goes in the ◯?

(a) 151 ◯ 138

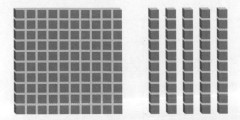

Hundreds	Tens	Ones
1	5	1

Hundreds	Tens	Ones
1	3	8

The hundreds are the same, so ...

(b) 852 ◯ 857

Hundreds	Tens	Ones
8	5	2

Hundreds	Tens	Ones
8	5	7

Both the hundreds and the tens are the same, so...

(c) 967 ◯ 796

Hundreds	Tens	Ones
9	6	7

Hundreds	Tens	Ones
7	9	6

1-6 Comparing Hundreds, Tens, and Ones

(d) 95  403

2 What sign, > or <, or =, goes in the 〇?

(a) 300 〇 700

(b) 798 〇 903

(c) 680 〇 95

(d) 227 〇 224

(e) 300 + 40 〇 340

(f) 708 〇 700 + 20

(g) 500 + 60 〇 300 + 70 + 8

(h) 40 + 700 + 1 〇 80 + 600 + 1

(i) eight hundred seventeen 〇 eight hundred sixty-seven

(j) two hundred eleven 〇 ninety-nine

3 (a) Which number is less, 892 or 798?

(b) Which number is greater, 98 or 401?

(c) Which number is the greatest, 670, 730, or 701?

(d) What is the greatest number that can be made using 4, 5, and 7?

(e) What is the least 3-digit number that can be made using 0, 8, and 3?

4 Arrange the numbers in order from least to greatest.

(a)
| 237 | 302 | 240 |

(b)
| 74 | 702 | 715 | 517 |

(c)
| 207 | 72 | 720 | 270 | 27 |

(d)
| 963 | 369 | 936 | 396 | 693 | 699 |

Exercise 6 • page 15

1-6 Comparing Hundreds, Tens, and Ones

Think

I have 586 stickers.

I have 100 more stickers than Sofia.

I have 200 fewer stickers than Sofia.

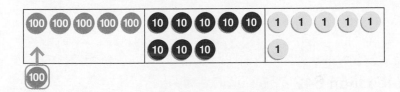

How many stickers do Alex and Mei have?

Learn

$586 + 100 =$ [ ]

Alex has [ ] stickers.

$586 - 200 =$ [ ]

Mei has [ ] stickers.

Do

1 (a) What number is 2 more than 387?

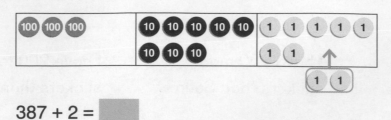

387 + 2 = ▢

(b) What number is 3 less than 205?

205 − 3 = ▢

(c) What number is 30 more than 435?

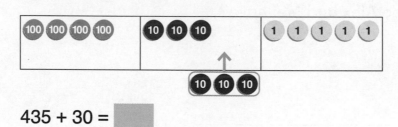

435 + 30 = ▢

(d) What number is 20 less than 642?

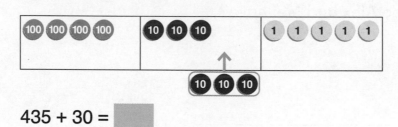

642 − 20 = ▢

2 (a) What number is 1 more than 339? $339 + 1 =$

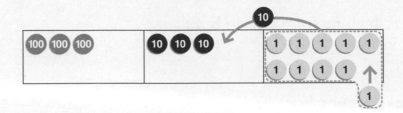

(b) What number is 1 less than 530? $530 - 1 =$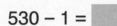

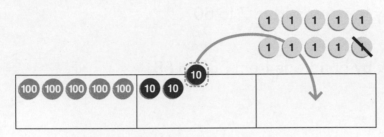

(c) What number is 10 more than 395? $395 + 10 =$

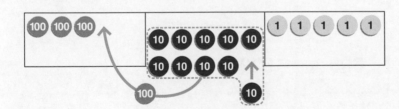

(d) What number is 10 less than 402? $402 - 10 =$

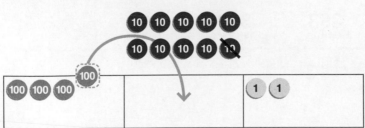

3 (a) Count on by ones from 346 to 354:
346, 347, 348, ..., 354.

(b) Count back by tens from 642 to 552:
642, 632, 622, ..., 552.

(c) Count back by ones from 874 to 865.

(d) Count on by tens from 156 to 246.

(e) Count on by hundreds from 7 to 607.

(f) Count back by hundreds from 589 to 89.

4 What are the missing numbers?

(a)
496	497	498	499		

(b)
604	603	602	601		

(c)
832	822	812	802		

(d)
394	393	392	391		

1 What number is shown?

(a)
| 100 100 100 100 100 | | 1 1 1 1 1
 1 1 1 |

(b)
| 100 100 100 100 100
 100 100 100 | 10 10 10 10 10
 10 | 1 1 |

(c)
| | 10 10 10 10 10
 10 10 | 1 1 1 1 |

(d)
| 100 100 100 100 100
 100 | 10 10 10 10 | |

2 Write the number in words.

(a) 611   (b) 321   (c) 450   (d) 405

3 Write the number.

(a) ninety-two   (b) three hundred eighty-seven

(c) 5 hundreds and 3 ones   (d) 1 one, 4 tens, and 2 hundreds

(e) 8 tens and 6 hundreds   (f) 3 hundreds, 6 tens, 17 ones

4 Put the numbers in order from greatest to least.

(a)

| 255 | 505 | 258 | 520 |

(b)

| 107 | 71 | 710 | 170 |

5 What sign, >, <, or =, goes in the ◯?

(a) 688 ◯ 742

(b) 570 ◯ 539

(c) 726 ◯ 729

(d) 300 + 7 ◯ 200 + 100 + 7

(e) 560 + 30 ◯ 560 + 7

(f) 3 hundreds ◯ 30 tens

6 What number is ...

(a) 2 more than 424

(b) 3 more than 767

(c) 2 less than 109

(d) 30 less than 482

(e) 20 less than 108

(f) 30 more than 680

7 (a) $145 - 2 =$ ▢

(b) $678 + 2 =$ ▢

(c) $874 + 20 =$ ▢

(d) $688 + 30 =$ ▢

(e) $500 - 20 =$ ▢

(f) $674 - 300 =$ ▢

Exercise 8 • page 23

1-8 Practice

Chapter 2

Addition and Subtraction — Part 1

Say the answers and look for patterns.

1 + 1	2 + 1	3 + 1	4 + 1	5 + 1	6 + 1	7 + 1	8 + 1	9 + 1
1 + 2	2 + 2	3 + 2	4 + 2	5 + 2	6 + 2	7 + 2	8 + 2	
1 + 3	2 + 3	3 + 3	4 + 3	5 + 3	6 + 3	7 + 3		
1 + 4	2 + 4	3 + 4	4 + 4	5 + 4	6 + 4			
1 + 5	2 + 5	3 + 5	4 + 5	5 + 5				
1 + 6	2 + 6	3 + 6	4 + 6					
1 + 7	2 + 7	3 + 7						
1 + 8	2 + 8							
1 + 9								

Which addition facts do you know quickly and easily?

Make flash cards for the facts you need to practice.

5 + 4	9
front	back

Say the answers and look for patterns.

2 – 1	3 – 1	4 – 1	5 – 1	6 – 1	7 – 1	8 – 1	9 – 1	10 – 1
	3 – 2	4 – 2	5 – 2	6 – 2	7 – 2	8 – 2	9 – 2	10 – 2
		4 – 3	5 – 3	6 – 3	7 – 3	8 – 3	9 – 3	10 – 3
			5 – 4	6 – 4	7 – 4	8 – 4	9 – 4	10 – 4
				6 – 5	7 – 5	8 – 5	9 – 5	10 – 5
					7 – 6	8 – 6	9 – 6	10 – 6
						8 – 7	9 – 7	10 – 7
							9 – 8	10 – 8
								10 – 9

Which subtraction facts do you know quickly and easily?

Make flash cards for the facts you need to practice.

9 – 6	3
front	back

Think

Mei has 8 guitar picks.

She buys 7 more guitar picks.

How many guitar picks does she have now?

Learn

Method 1

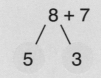

8 + 7
/ \
2   5

Method 2

8 + 7 = [ ]

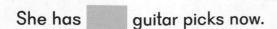

She has [ ] guitar picks now.

8 + 7
/ \
5   3

Do

1 (a) Add 7 and 6.

7 + 6 = [ ]

7 + 6
/ \
3   3

(b) Add 5 and 9.

5 + 9 = [ ]

5 + 9
/ \
4   1

2 (a) 9 + 3 = [ ]   (b) 7 + 5 = [ ]   (c) 5 + 8 = [ ]

(d) 7 + 7 = [ ]   (e) 6 + 5 = [ ]   (f) [ ] = 8 + 8

3 Look for patterns.

Make flash cards for the
facts you need to practice.

3 + 9	12
front	back

							9 + 2
						8 + 3	9 + 3
					7 + 4	8 + 4	9 + 4
				6 + 5	7 + 5	8 + 5	9 + 5
			5 + 6	6 + 6	7 + 6	8 + 6	9 + 6
		4 + 7	5 + 7	6 + 7	7 + 7	8 + 7	9 + 7
	3 + 8	4 + 8	5 + 8	6 + 8	7 + 8	8 + 8	9 + 8
2 + 9	3 + 9	4 + 9	5 + 9	6 + 9	7 + 9	8 + 9	9 + 9

Exercise 1 • page 27

2-1 Strategies for Addition

Think

Alex has 15 cards.
He gives away 7 of them.
How many cards does he have left?

Learn

Method 1

$$15 - 7$$
$$5 \qquad 10$$

Method 2

$$15 - 7$$
$$5 \qquad 2$$

$15 - 7 = $ ▢

He has ▢ cards left.

Do

1 (a) Subtract 9 from 13.

$13 - 9 = $ ▢

(b) Subtract 8 from 14.

$14 - 8 = $ ▢

2 (a) $16 - 9 = $ ▢ (b) $15 - 9 = $ ▢ (c) $14 - 5 = $ ▢

(d) $11 - 7 = $ ▢ (e) $13 - 8 = $ ▢ (f) ▢ $= 12 - 9$

3 Look for patterns.

Make flash cards for the facts you need to practice.

13 − 6	7
front	back

11 − 2

| 11 − 3 | 12 − 3 |

| 11 − 4 | 12 − 4 | 13 − 4 |

| 11 − 5 | 12 − 5 | 13 − 5 | 14 − 5 |

| 11 − 6 | 12 − 6 | 13 − 6 | 14 − 6 | 15 − 6 |

| 11 − 7 | 12 − 7 | 13 − 7 | 14 − 7 | 15 − 7 | 16 − 7 |

| 11 − 8 | 12 − 8 | 13 − 8 | 14 − 8 | 15 − 8 | 16 − 8 | 17 − 8 |

| 11 − 9 | 12 − 9 | 13 − 9 | 14 − 9 | 15 − 9 | 16 − 9 | 17 − 9 | 18 − 9 |

Exercise 2 • page 29

Lesson 3
Parts and Whole

Think

Do we add or subtract?

Explain why.

(a) Emma has 8 comic books.

Dion has 6 comic books.

How many comic books do they have altogether?

(b) Emma and Dion have 14 comic books altogether.

Emma has 8 comic books.

How many comic books does Dion have?

Learn

(a)

Total number of comic books

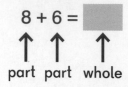

?

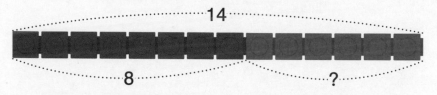

8

Emma's comic books

6

Dion's comic books

$8 + 6 =$

↑ part ↑ part ↑ whole

They have ▢ comic books altogether.

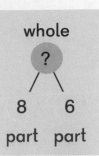

whole

?

8 ⎯ part

6 ⎯ part

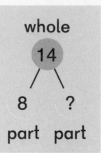

(b)

Total number of comic books

14

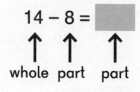

8

Emma's comic books

?

Dion's comic books

$14 - 8 =$

↑ whole ↑ part ↑ part

Dion has ▢ comic books.

whole

14

8 ⎯ part

? ⎯ part

Do

1 (a)

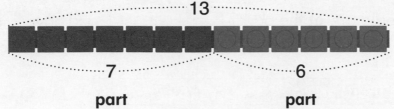

$7 + 6 = \boxed{}$ $6 + 7 = \boxed{}$

$13 - 7 = \boxed{}$ $13 - 6 = \boxed{}$

(b)

16 — 7
16 — 9

$9 + 7 = \boxed{}$ $7 + 9 = \boxed{}$

$16 - 9 = \boxed{}$ $16 - 7 = \boxed{}$

(c)

4 — 12
8 — 12

$8 + \boxed{} = 12$ $\boxed{} + 8 = 12$

$12 - \boxed{} = 8$ $12 - 8 = \boxed{}$

2 There are 14 children at the park.
6 children go home.
How many children are still at the park?

> We know the whole and one part, so we subtract.

14	
still at park	gone home
?	6

$\boxed{} \bigcirc \boxed{} = \boxed{}$

14 — ? — 6

There are $\boxed{}$ children still at the park.

3 There are 11 children and 8 adults at the park.
How many people are at the park?

We know two parts and have to find the whole, so we add.

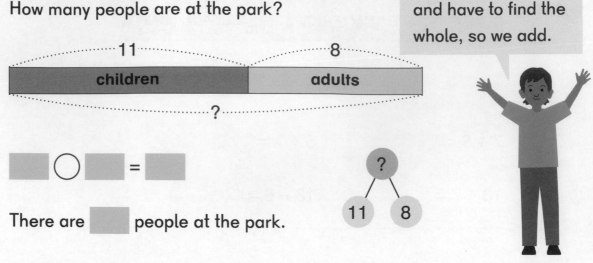

There are ☐ people at the park.

4 Write an equation and find the answer.

(a) There are 17 children at the park.
Some of the children go home.
8 children are still playing at the park.
How many went home?

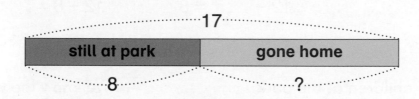

(b) There were 7 children at the park.
Some more children came.
Now there are 10 children at the park altogether.
How many more children came?

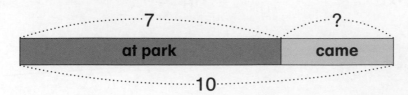

Exercise 3 • page 33

Think

There are 12 ants and 7 crickets
near a picnic table.

(a) How many more ants than
crickets are there?
Write an equation.

(b) How many insects are there altogether?
Write an equation.

Learn

12

ants

crickets

7

?

?

(a) $12 - 7 =$ ☐

There are ☐ more ants than crickets.

(b) $12 + 7 =$ ☐

There are ☐ insects altogether.

Subtract to find
the difference.

Do

1 There are 11 ducks and 5 swans in a pond.

(a) How many ducks and swans are there altogether?

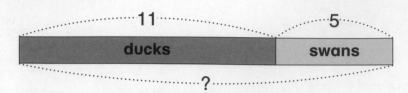

11   5

ducks   swans

?

[ ] ◯ [ ] = [ ]

There are [ ] ducks and swans altogether.

(b) How many fewer swans than ducks are there?

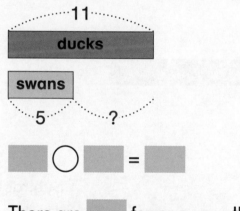

11

ducks

swans

5   ?

[ ] ◯ [ ] = [ ]

There are [ ] fewer swans than ducks.

2

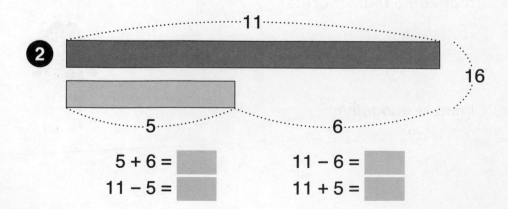

11

16

5   6

5 + 6 = [ ]   11 − 6 = [ ]
11 − 5 = [ ]   11 + 5 = [ ]

Write an equation and find the answer.

3 Kaylee is 7 years old.
Her sister Madison is 4 years older than Kaylee.
How old is Madison?

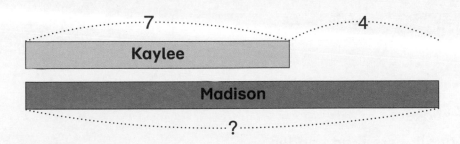

4 Hudson has 18 stickers.
He has 5 more stickers than Darryl.
How many stickers does Darryl have?

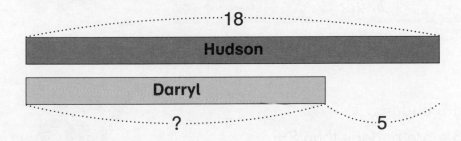

5 Jordon has 10 sports cards.
He has 4 fewer sports cards than Mila.

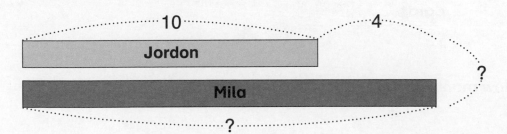

(a) How many cards does
Mila have?

(b) How many cards do they
have altogether?

6 There were some cherries in a bag.
Natasha ate 6 of them.
Now there are 8 cherries left.
How many cherries were there at first?

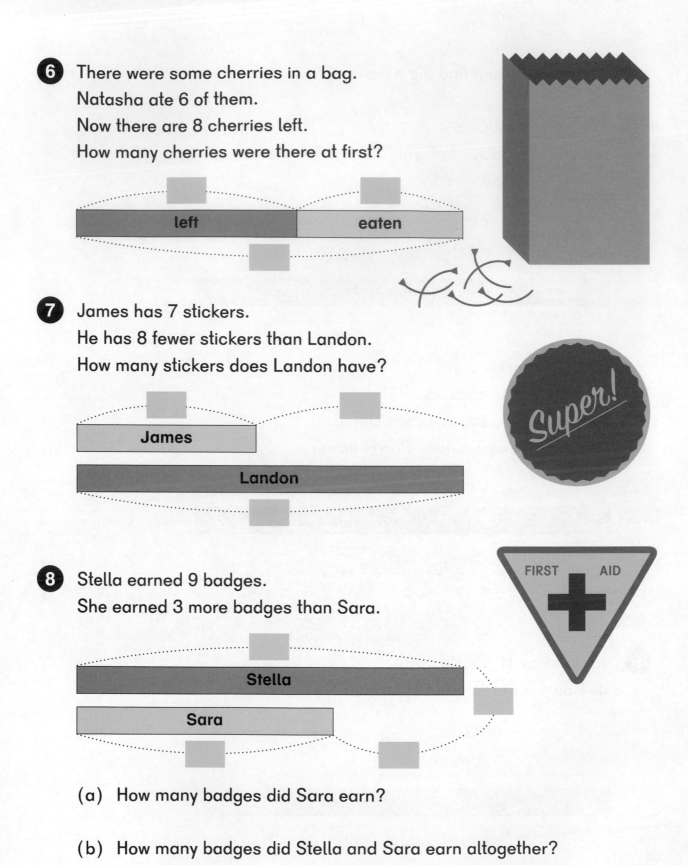

7 James has 7 stickers.
He has 8 fewer stickers than Landon.
How many stickers does Landon have?

8 Stella earned 9 badges.
She earned 3 more badges than Sara.

(a) How many badges did Sara earn?

(b) How many badges did Stella and Sara earn altogether?

Exercise 4 • page 37

2-4 Comparison

1 (a) 6 + 8 = ▢   (b) 5 + 7 = ▢

(c) 9 + 6 = ▢   (d) 8 + 5 = ▢

(e) 17 − 9 = ▢   (f) 12 − 8 = ▢

(g) 13 − 4 = ▢   (h) ▢ + 8 = 14

(i) 4 + ▢ = 11   (j) 14 − ▢ = 7

(k) ▢ − 6 = 8   (l) 14 + 3 = ▢

(m) 12 + 5 = ▢   (n) 19 − 2 = ▢

(o) 16 − 4 = ▢   (p) 12 + 7 = ▢

(q) ▢ = 11 + 8   (r) 8 = 11 − ▢

2 Julian had 4 sports cards.
He bought 6 more sports cards.
How many sports cards does he have now?

3 There are 13 ladybugs in all.
5 ladybugs are on top of the leaf.
The rest are hidden under the leaf.
How many ladybugs are hidden?

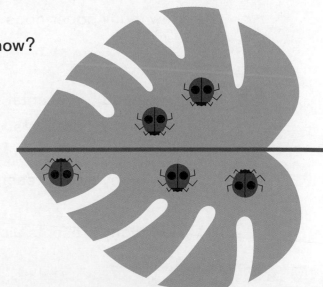

4 Valentina needs 20 beads for her necklace.
She has 8 beads.
How many more beads does she need?

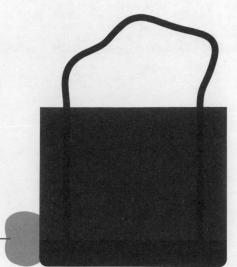

5 There are 13 apples.
5 of them are in a bag.
How many apples are not in the bag?

6 There are 9 flowers in a blue vase.
There are 3 fewer flowers in a red vase than in the blue vase.

(a) How many flowers does the red vase have?

(b) How many flowers are in both vases?

7 Jett made 11 paper dogs.
He made 6 more paper dogs than his sister.

(a) How many paper dogs did his sister make?

(b) How many paper dogs did they make altogether?

8 Kawai wrapped 5 presents.
Nina wrapped 12 presents.

(a) How many more presents did Nina wrap than Kawai?

(b) How many presents did they wrap in all?

Exercise 5 • page 41

2-5 Practice

Chapter 3

Addition and Subtraction — Part 2

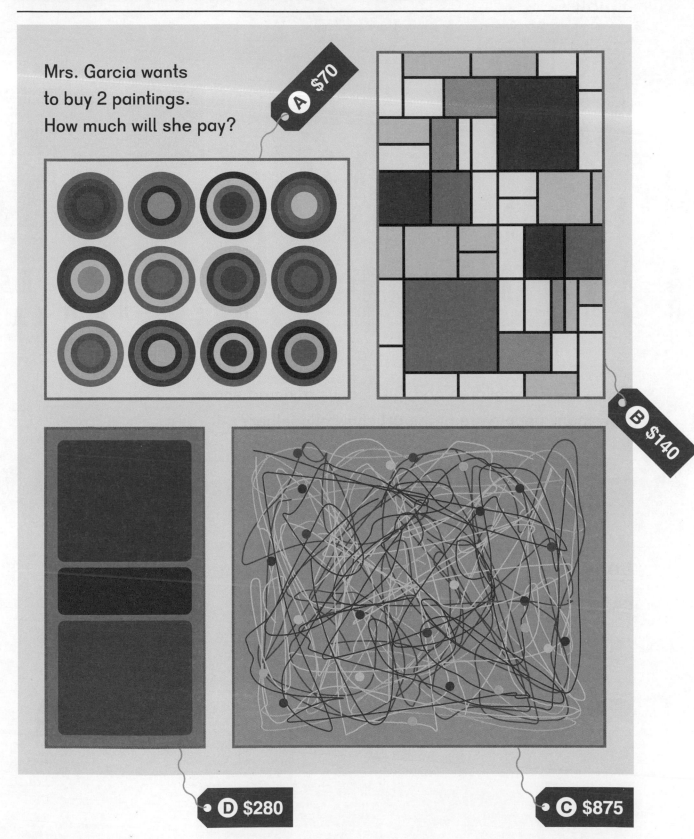

Mrs. Garcia wants to buy 2 paintings. How much will she pay?

A $70

B $140

C $875

D $280

Think

There are 256 children and 341 adults at an art show.

How many people are at the art show altogether?

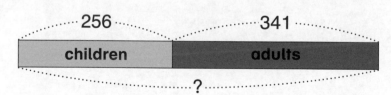

Learn

256 + 341 = ?

To find the whole, we add.

Add the ones.

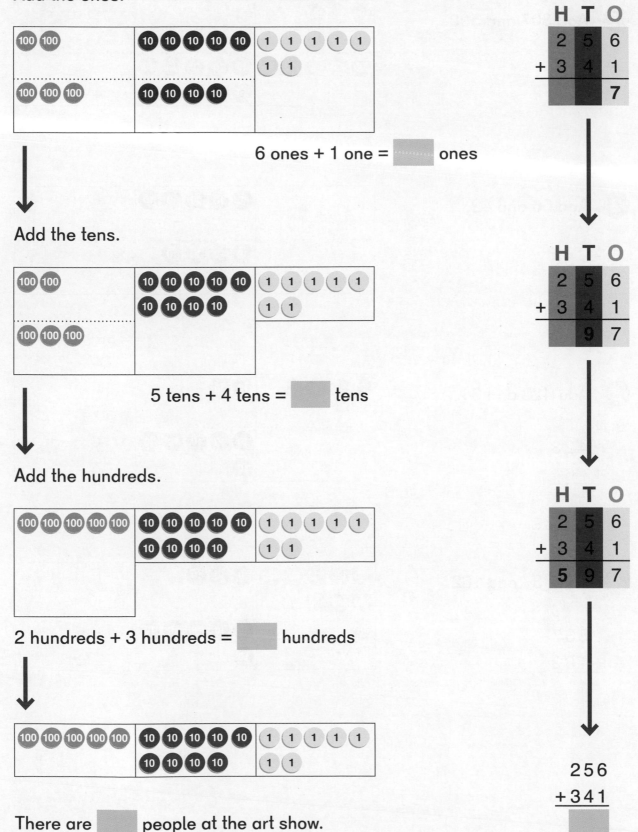

6 ones + 1 one = ones

Add the tens.

5 tens + 4 tens = tens

Add the hundreds.

2 hundreds + 3 hundreds = hundreds

There are people at the art show.

256
+341

Do

1 Add 307 and 590.

307
+590
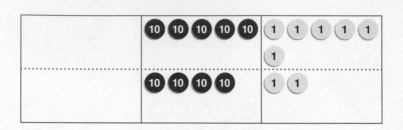

2 Add 56 and 42.

56
+ 42
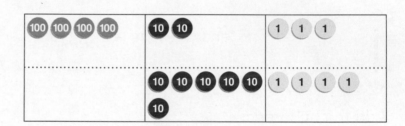

3 Add 423 and 64.

423
+ 64
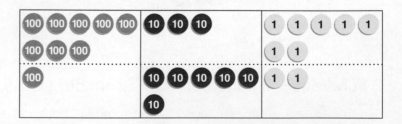

4 Add 837 and 162.

837
+162

5 Find the value.

(a) 61 + 8

(b) 43 + 24

(c) 501 + 6

(d) 326 + 43

(e) 525 + 162

(f) 202 + 330

(g) 451 + 246

(h) 736 + 152

(i) 505 + 310

(j) 55 + 213

6 There are 352 bees in a beehive.
47 more bees come to the hive.
How many bees are in the beehive now?

There are ▢ bees in the beehive now.

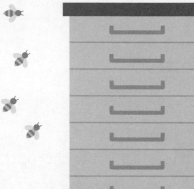

7 Eliza has 456 stamps.
She has 122 fewer stamps than Claudia.
How many stamps does Claudia have?

▢ ◯ ▢ = ▢

Claudia has ▢ stamps.

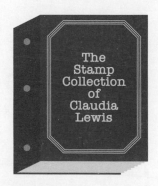

Exercise 1 • page 45

Think

There are 584 children in 2 leagues.
253 of them are in the Cheetah League.
The rest are in the Roadrunner League.
How many children are in
the Roadrunner League?

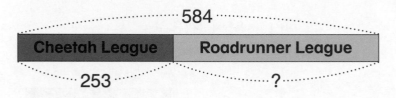

Learn

$584 - 253 = ?$

To find a part,
we subtract.

Subtract the ones.

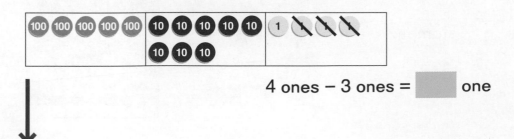

4 ones − 3 ones = ⬜ one

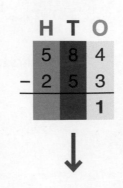

Subtract the tens.

$$8 \text{ tens} - 5 \text{ tens} = \boxed{} \text{ tens}$$

Subtract the hundreds.

$$5 \text{ hundreds} - 2 \text{ hundreds} = \boxed{} \text{ hundreds}$$

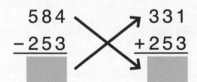

There are $\boxed{}$ children in the Roadrunner League.

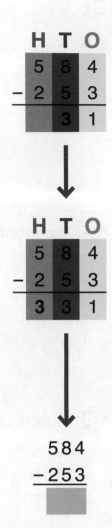

H	T	O
5	8	4
− 2	5	3
	3	1

H	T	O
5	8	4
− 2	5	3
3	3	1

$$584$$
$$-253$$
$$\boxed{}$$

$$\begin{array}{c} 584 \\ -253 \\ \hline \boxed{} \end{array} \quad\times\quad \begin{array}{c} 331 \\ +253 \\ \hline \boxed{} \end{array}$$

Add to check the answer.

Do

1 Subtract 404 from 678.

```
  678
- 404
```

2 Subtract 41 from 76.

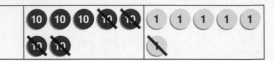

```
   76
-  41
```

3 Subtract 29 from 789.

```
  789
-  29
```

4 Subtract 125 from 569.

```
  569
- 125
```

5 Find the value.

(a) 89 – 7

(b) 65 – 43

(c) 739 – 7

(d) 156 – 45

(e) 683 – 271

(f) 607 – 503

6 What are the missing digits?

(a)
```
    9 [ ] 8
  – 4   3 [ ]
  ─────────
    5   3   8
```

(b)
```
    3   8 [ ]
  – [ ] 3   5
  ─────────
      [ ] 2
```

7 There are 387 sunflower seeds.
Some birds eat 125 seeds.
How many seeds are left?

[ ] ◯ [ ] = [ ]

[ ] seeds are left.

8 There are 179 sparrows eating the seeds.
There are 121 fewer robins than sparrows.
How many robins are there?

[ ] ◯ [ ] = [ ]

There are [ ] robins.

Exercise 2 • page 49

Lesson 3
Addition with Regrouping Ones

Think

Dion collected 158 box tops.
Sofia collected 136 box tops.
How many box tops did they collect altogether?

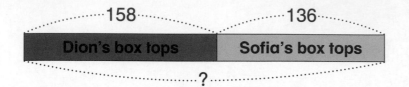

158   136

| Dion's box tops | Sofia's box tops |

?

BOX TOPS

Learn

158 + 136 = ?

There are more than 10 ones.

Add the ones.

8 ones + 6 ones = 14 ones

Regroup the ones.

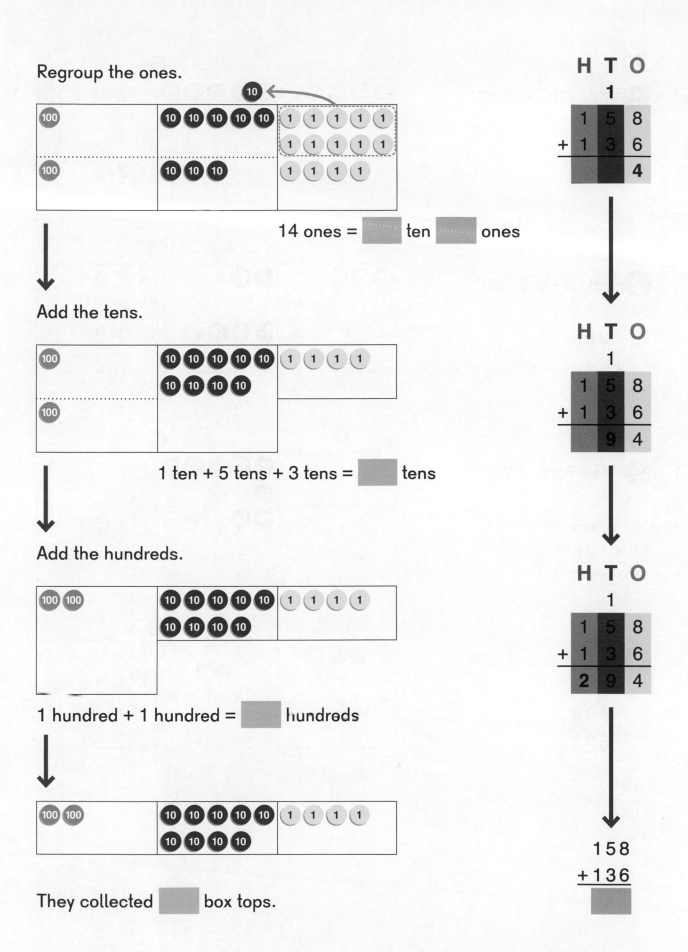

14 ones = ☐ ten ☐ ones

Add the tens.

1 ten + 5 tens + 3 tens = ☐ tens

Add the hundreds.

1 hundred + 1 hundred = ☐ hundreds

They collected ☐ box tops.

Do

1 Add 685 and 207.

685
+207
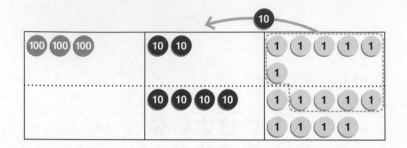

2 Add 49 and 326.

326
+ 49

3 Add 65 and 25.

65
+ 25

4 Add 408 and 507.

408
+507

5 Find the value.

(a) 45 + 7

(b) 38 + 48

(c) 609 + 8

(d) 78 + 302

(e) 175 + 209

(f) 746 + 234

6 Explain the errors made in the calculations below
and find the correct answer.

(a)
```
   43
+  38
  711
```
✗

(b)
```
   43
+  38
   71
```
✗

7 2 years ago, there were 319 members in a club.
Last year, 109 new members joined.
This year, 65 new members have joined so far.

(a) At the end of last year, how many members did the club have?

◻ ◯ ◻ = ◻

The club had ◻ members at the end of last year.

(b) How many members does the club have now?

◻ ◯ ◻ = ◻

The club now has ◻ members.

Exercise 3 • page 53

Lesson 4
Addition with Regrouping Tens

④

Think

463 children and 375 adults went to the zoo on Saturday.
How many people went to the zoo altogether?

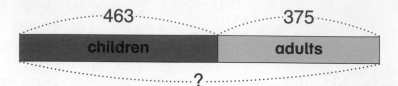

Learn

463 + 375 = ?

There are more than 10 tens.

Add the ones.

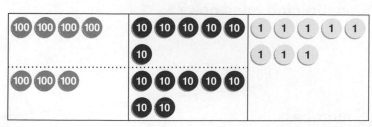

3 ones + 5 ones = ☐ ones

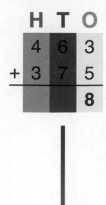

H	T	O
4	6	3
+ 3	7	5
		8

Add the tens.

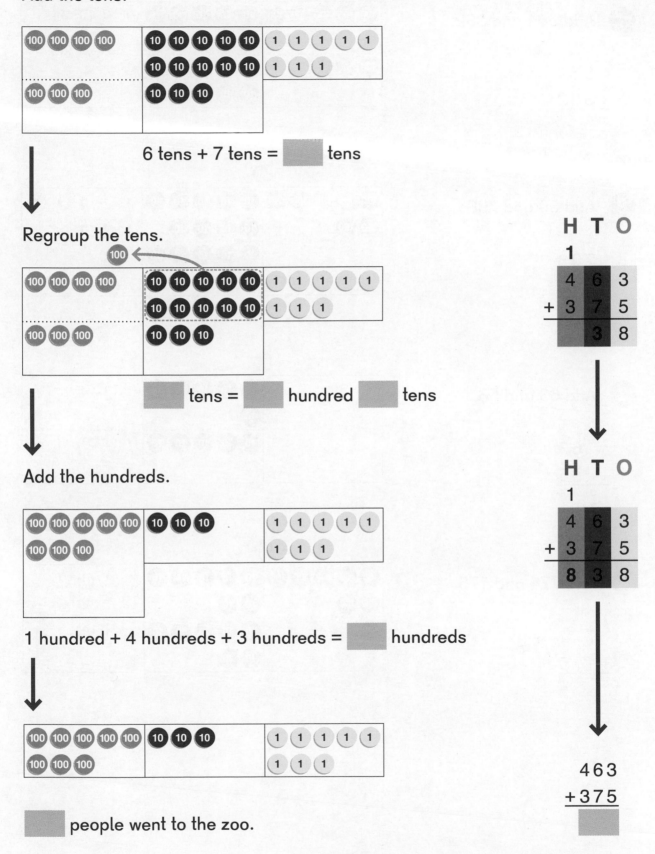

6 tens + 7 tens = ▢ tens

Regroup the tens.

▢ tens = ▢ hundred ▢ tens

Add the hundreds.

1 hundred + 4 hundreds + 3 hundreds = ▢ hundreds

▢ people went to the zoo.

H T O
1
4 6 3
+ 3 7 5
    3 8

H T O
1
4 6 3
+ 3 7 5
8 3 8

    4 6 3
+ 3 7 5
    ▢

Do

1 Add 584 and 382.

$$\begin{array}{r} 584 \\ +382 \\ \hline \end{array}$$

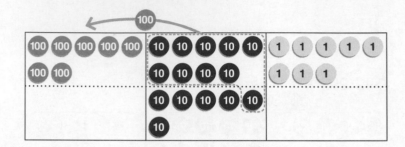

2 Add 60 and 798.

$$\begin{array}{r} 798 \\ +60 \\ \hline \end{array}$$

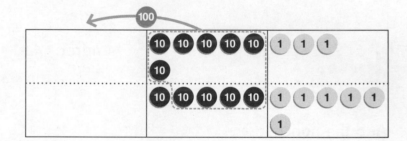

3 Add 63 and 56.

$$\begin{array}{r} 63 \\ +56 \\ \hline \end{array}$$

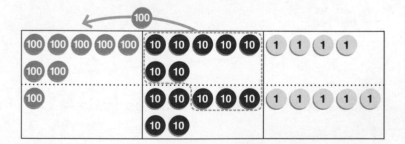

4 Add 774 and 175.

$$\begin{array}{r} 774 \\ +175 \\ \hline \end{array}$$

5 Find the value.

(a) 85 + 72

(b) 673 + 50

(c) 77 + 380

(d) 294 + 351

(e) 480 + 223

(f) 637 + 272

6 What are the missing digits?

(a)
```
      8 ▨
  +   3 6
  ▨ 1 8
```

(b)
```
      4 3 1
  + 4 ▨ 5
  ▨ 0 ▨
```

7 There are 186 coins in a jar.
Ethan put in 243 more coins.
How many coins are in the jar now?

Ethan's Bicycle Savings

There are ☐ coins in the jar now.

8 A pine tree is 95 years old.
A redwood tree is 832 years older than the pine tree.
How old is the redwood tree?

The redwood tree is ☐ years old.

Exercise 4 • page 57

Think

Sofia has 237 red beads and 385 blue beads to make bracelets.
How many beads does she have in all?

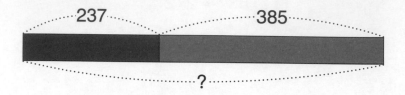

Learn

$237 + 385 = ?$

There are more than 10 ones
and more than 10 tens.

Add the ones.

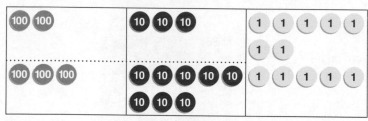

7 ones + 5 ones = ☐ ones

Regroup the ones.

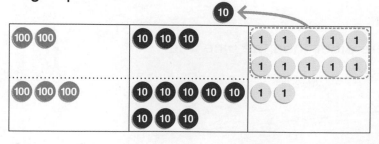

12 ones = 1 ten 2 ones

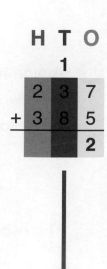

	H	T	O
			1
	2	3	7
+	3	8	5
			2

Add the tens.

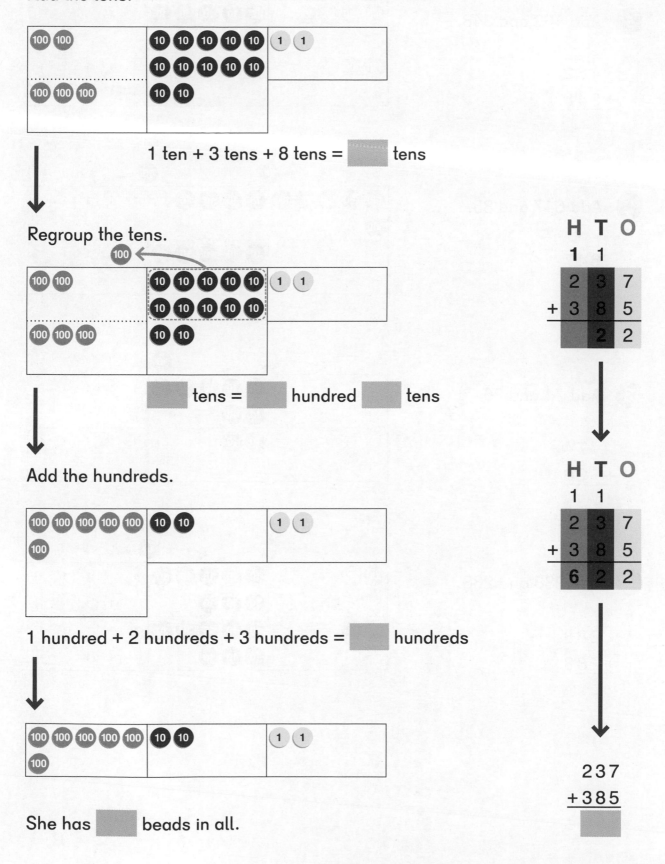

1 ten + 3 tens + 8 tens = ▢ tens

Regroup the tens.

▢ tens = ▢ hundred ▢ tens

Add the hundreds.

1 hundred + 2 hundreds + 3 hundreds = ▢ hundreds

She has ▢ beads in all.

H T O
1 1
2 3 7
+ 3 8 5
    2 2

H T O
1 1
2 3 7
+ 3 8 5
6 2 2

  2 3 7
+ 3 8 5
▢

Do

1 Add 462 and 348.

```
  462
+ 348
```
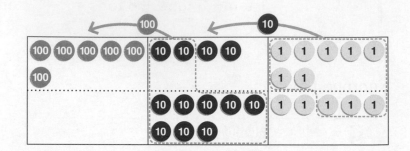

2 Add 647 and 85.

```
  647
+  85
```
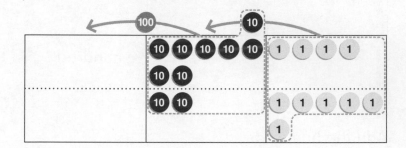

3 Add 74 and 26.

```
   74
+  26
```
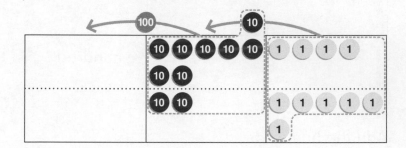

4 Add 288 and 288.

```
  288
+ 288
```
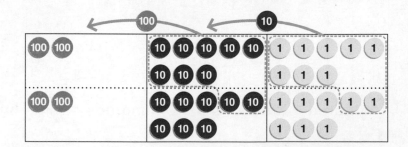

5 Find the value.

(a) 62 + 49

(b) 85 + 119

(c) 365 + 287

(d) 544 + 276

(e) 457 + 348

(f) 399 + 399

(g) 733 + 167

(h) 285 + 159

6 John has $259.
Samuel has $192 more than John.

(a) How much money does Samuel have?

  [ ] ◯ [ ] = [ ]

  Samuel has $[   ].

(b) How much money do they have altogether?

  [ ] ◯ [ ] = [ ]

  They have $[   ] altogether.

7 Catalina has $100.
She wants to buy shoes for $65 and a shirt for $37.
Does she have enough money?
If not, how much more money does she need?

Exercise 5 • page 61

❶ Find the value.

(a) 33 + 5   (b) 67 − 5

(c) 89 − 42   (d) 179 − 60

(e) 356 + 23   (f) 183 − 61

(g) 689 − 427   (h) 423 + 54

(i) 608 + 191   (j) 305 − 102

❷ A farmer had 580 tomatoes.
He sold 430 of them and used the rest to make tomato sauce.
How many tomatoes did he use to make the sauce?

❸ A photograph costs $110 and a painting costs $280.
How much more does the painting cost than the photograph?

❹ Fadiya wants to buy a coat and some shoes.
The coat costs $153 and the shoes cost $32.

(a) How much more does the coat cost than the shoes?

(b) How much money does Fadiya need to buy both items?

5 Find the value.

(a) 67 + 18

(b) 85 + 63

(c) 764 + 77

(d) 256 + 37

(e) 519 + 290

(f) 483 + 139

(g) 289 + 164

(h) 132 + 379

6 What digits are missing?

(a)
```
    9  6  ▨
  +    2  6
  ▨ ▨  3
```

(b)
```
    3  8  ▨
  + 5  ▨  5
  ▨  2  3
```

7 Mrs. Garcia bought these two paintings.
How much did she spend?

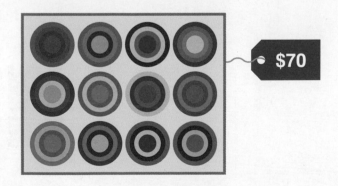
$70

$280

8 At the school fair, the second graders sold 235 cupcakes in the morning and 283 cupcakes in the afternoon.
How many cupcakes did they sell in all?

Exercise 6 • page 65

Lesson 7
Subtraction with Regrouping from Tens

Think

There are 262 people at a festival.

148 of them are children.

How many adults are there?

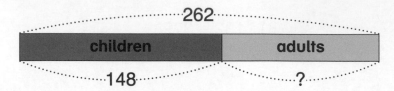

262

| children | adults |

148   ?

Learn

262 – 148 = ?

I cannot take 8 ones
away from the 2 ones.

Regroup 1 ten.

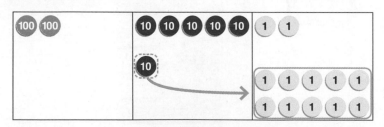

6 tens 2 ones = 5 tens 12 ones

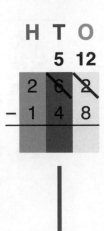

H	T	O
	5	12
2	6̶	2̶
– 1	4	8

Subtract the ones.

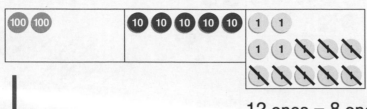

12 ones − 8 ones = ☐ ones

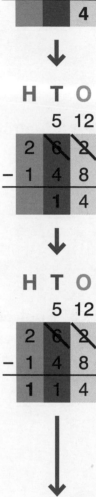

Subtract the tens.

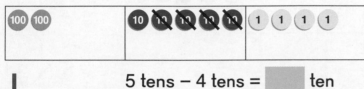

5 tens − 4 tens = ☐ ten

Subtract the hundreds.

2 hundreds − 1 hundred = ☐ hundred

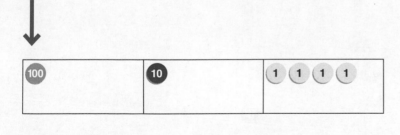

There are ☐ adults at the festival.

$$262 - 148$$

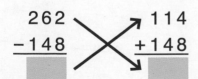

114
+148

Add to check the answer.

Do

1 Subtract 129 from 734.

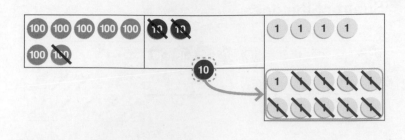

```
  734
- 129
```
▢

2 Subtract 43 from 72.

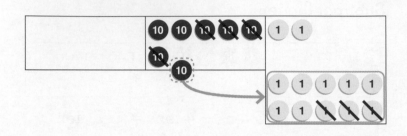

```
   72
-  43
```
▢

3 Subtract 56 from 880.

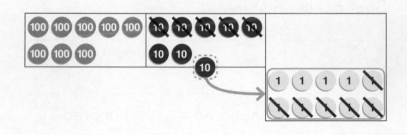

```
  880
-  56
```
▢

4 Subtract 307 from 516.

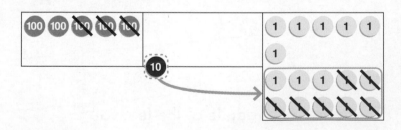

```
  516
- 307
```
▢

5 Find the value.

(a) 31 − 8

(b) 53 − 35

(c) 40 − 17

(d) 282 − 7

(e) 370 − 67

(f) 561 − 29

(g) 410 − 308

(h) 772 − 234

(i) 390 − 143

(j) 920 − 413

6 What are the missing digits?

(a)
```
    8 □
−   □ 6
────────
    3 5
```

(b)
```
  □ 3 2
+ 2 □ 5
────────
  5 0 □
```

7 Tomas wants to buy a camera that costs $560.
He has $423.
How much more money does he need?

$560

□ ◯ □ = □

He needs $ □ more.

Exercise 7 • page 67

Think

There are 538 red and blue beads in a jar.

263 of them are red beads.

How many blue beads are there?

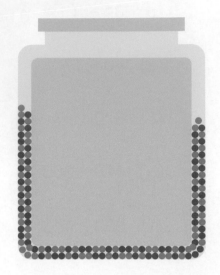

Learn

538 − 263 = ?

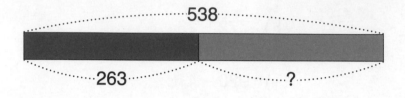

Subtract the ones.

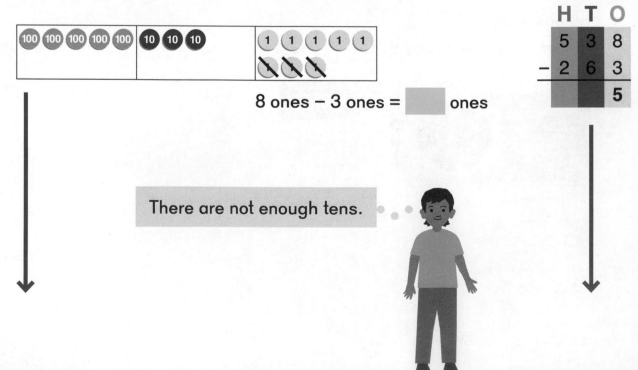

8 ones − 3 ones =   ones

	H	T	O
	5	3	8
−	2	6	3
			5

There are not enough tens.

Regroup 1 hundred.

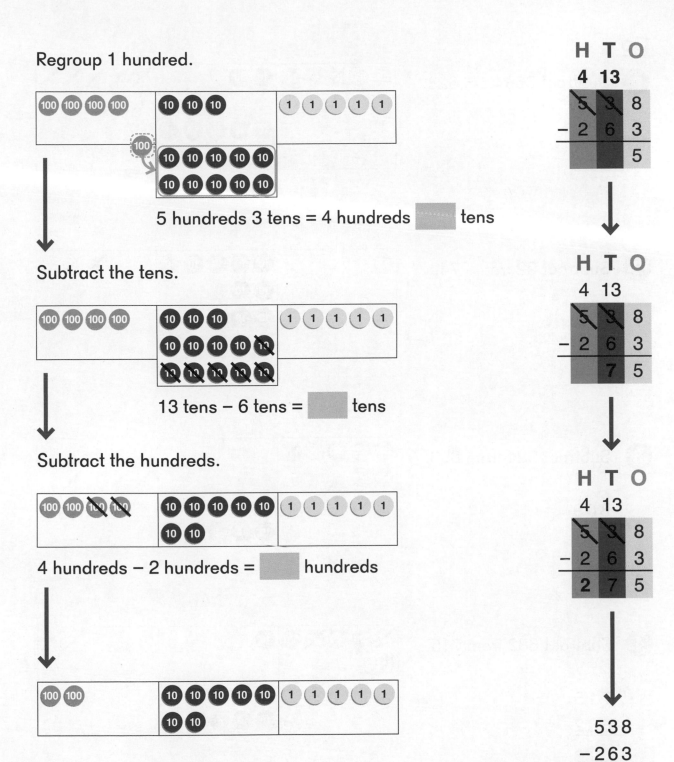

5 hundreds 3 tens = 4 hundreds ⬚ tens

Subtract the tens.

13 tens − 6 tens = ⬚ tens

Subtract the hundreds.

4 hundreds − 2 hundreds = ⬚ hundreds

There are ⬚ blue beads.

	H	T	O
	4	13	
	5̶	3̶	8
−	2	6	3
			5

	H	T	O
	4	13	
	5̶	3̶	8
−	2	6	3
		7	5

	H	T	O
	4	13	
	5̶	3̶	8
−	2	6	3
	2	7	5

$$\begin{array}{r} 538 \\ -263 \\ \hline \end{array}$$

⬚

Do

1 Subtract 356 from 628.

$$\begin{array}{r} 628 \\ -356 \\ \hline \end{array}$$

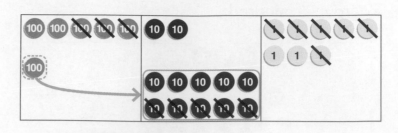

2 Subtract 92 from 274.

$$\begin{array}{r} 274 \\ -92 \\ \hline \end{array}$$

3 Subtract 324 from 809.

$$\begin{array}{r} 809 \\ -324 \\ \hline \end{array}$$

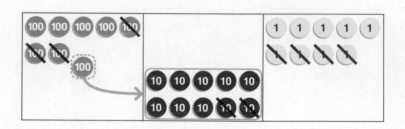

4 Subtract 682 from 715.

$$\begin{array}{r} 715 \\ -682 \\ \hline \end{array}$$

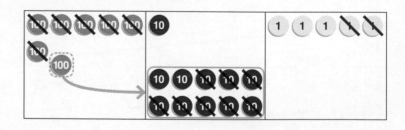

5 Find the value.

(a) 256 − 75

(b) 408 − 62

(c) 783 − 391

(d) 607 − 345

(e) 419 − 199

(f) 727 − 634

6 Kona has $444.
She bought a tennis racket for $183.
How much money does she have left?

She has $ ⬜ left.

7 Susma collected 137 seashells.
Anna collected 93 seashells.

(a) How many more seashells did Susma collect than Anna?

Susma collected ⬜ more seashells than Anna.

(b) How many seashells did they collect altogether?

They collected ⬜ seashells altogether.

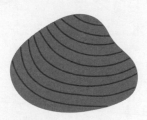

Exercise 8 • page 71

Lesson 9
Subtraction with Regrouping from Two Places

Think

There are 344 animals at the dog and cat hospital.
179 of them are dogs.
How many cats are there?

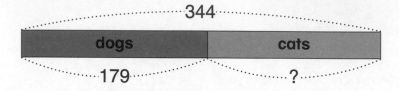

Learn

344 – 179 = ?

Regroup 1 ten.

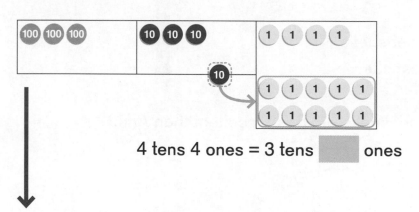

4 tens 4 ones = 3 tens ▢ ones

Subtract the ones.

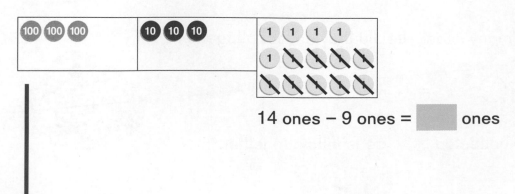

14 ones – 9 ones = ▢ ones

Regroup 1 hundred.

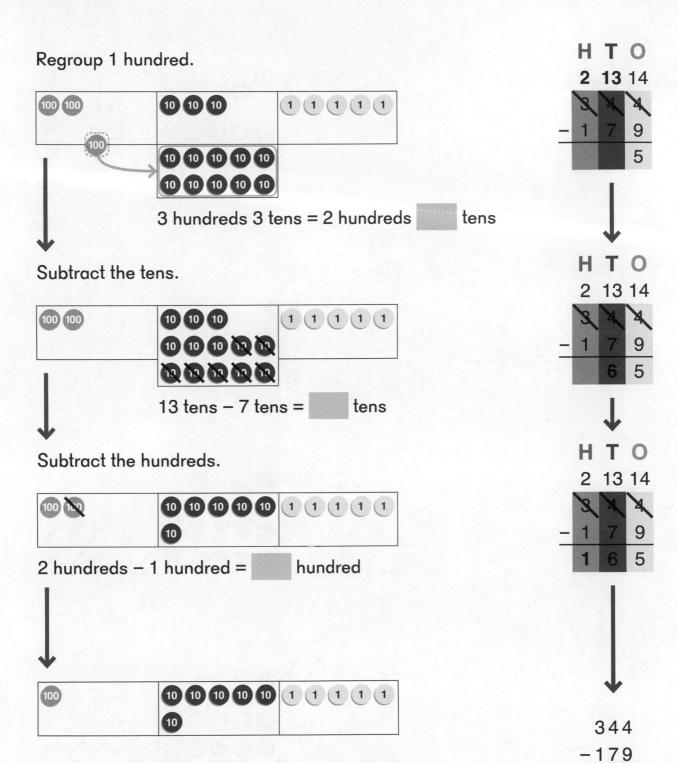

3 hundreds 3 tens = 2 hundreds ▨ tens

Subtract the tens.

13 tens − 7 tens = ▨ tens

Subtract the hundreds.

2 hundreds − 1 hundred = ▨ hundred

There are ▨ cats.

H T O
2 13 14
   3̶ 4̶ 4̶
− 1 7 9
         5

H T O
2 13 14
3̶ 4̶ 4̶
− 1 7 9
      6 5

H T O
2 13 14
3̶ 4̶ 4̶
− 1 7 9
   1 6 5

  3 4 4
− 1 7 9
  ▨

Do

1 Subtract 386 from 620.

$$\begin{array}{r} 620 \\ -386 \\ \hline \end{array}$$

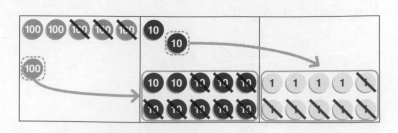

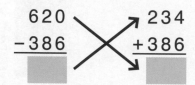

Check the answer using addition.

$$\begin{array}{r} 620 \\ -386 \\ \hline \end{array} \qquad \begin{array}{r} 234 \\ +386 \\ \hline \end{array}$$

2 Subtract 89 from 523.

$$\begin{array}{r} 523 \\ -\ 89 \\ \hline \end{array}$$

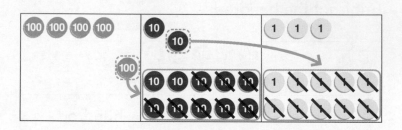

3 Subtract 537 from 612.

$$\begin{array}{r} 612 \\ -537 \\ \hline \end{array}$$

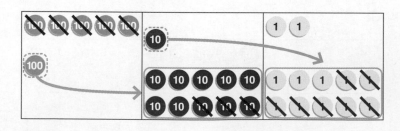

4 Subtract 44 from 123.

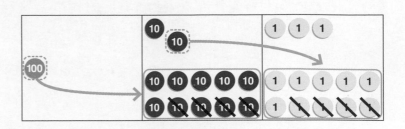

```
   1 2 3
 −   4 4
 _____
```

5 Find the value.

(a) 348 − 69

(b) 450 − 73

(c) 562 − 294

(d) 760 − 381

(e) 613 − 237

(f) 810 − 184

6 Boris had $450.
He bought a jacket and a pair of shoes.
He has $116 left.

(a) How much did he spend?

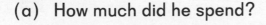

He spent $ ___ .

(b) The shoes cost $98.
How much did the jacket cost?

The jacket cost $ ___ .

Exercise 9 · page 75

Think

A store had 403 tomatoes.

185 of the tomatoes were sold.

How many tomatoes are left?

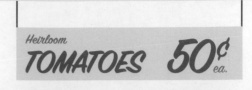

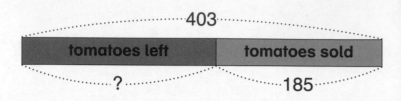

Learn

403 − 185 = ?

I need more ones,
but there are no tens to regroup.

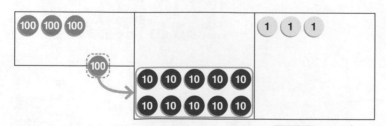

Regroup 1 hundred.

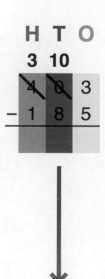

4 hundreds 0 tens = 3 hundreds ▢ tens

Regroup 1 ten.

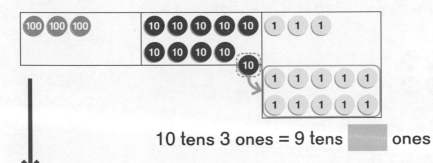

10 tens 3 ones = 9 tens ___ ones

Subtract the ones.

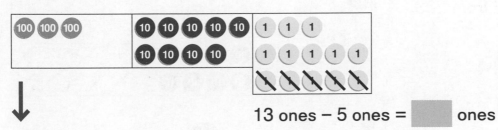

13 ones − 5 ones = ___ ones

Subtract the tens.

9 tens − 8 tens = ___ ten

Subtract the hundreds.

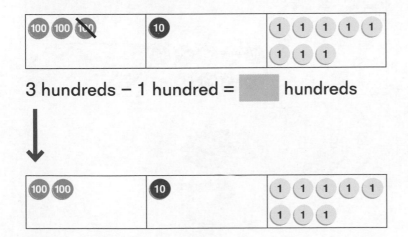

3 hundreds − 1 hundred = ___ hundreds

There are ___ tomatoes left.

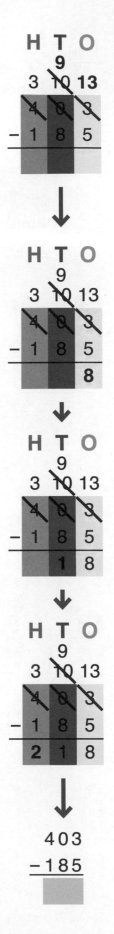

Do

1 1 hundred = ▢ tens   1 ten = ▢ ones

1 hundred = 9 tens and ▢ ones

2 Subtract 56 from 203.

$$\begin{array}{r} 203 \\ -\ 56 \\ \hline \end{array}$$

3 Subtract 267 from 500.

$$\begin{array}{r} 500 \\ -267 \\ \hline \end{array}$$

4 Find the value.

4 tens + ? tens = 9 tens

5 ones + ? ones = 10 ones

(a) 100 – 45   (b) 400 – 45

(c) 100 – 77   (d) 700 – 77

5 Find the value.

(a) 607 – 79   (b) 500 – 65

(c) 304 – 80   (d) 701 – 235

(e) 600 – 239   (f) 402 – 147

6 Explain the error made in the calculation
and find the correct answer.

```
      6  10 10
      7̶  0̶  0̶
   -  1  8  5
   ─────────────
      5  2  5  ✗
```

7 Ximena saved $157 last year.
She saved some more money this year.
She has now saved $304 altogether.
How much money did she save this year?

⬜ ◯ ⬜ = ⬜

She saved $⬜ this year.

1 Find the value.

(a) 81 – 43

(b) 70 – 57

(c) 285 – 78

(d) 340 – 7

(e) 460 – 88

(f) 485 – 329

(g) 706 – 535

(h) 615 – 276

(i) 707 – 329

(j) 200 – 102

(k) 700 – 312

(l) 902 – 289

2 There are 762 people at a sporting event.
251 are children.
How many adults are there?

3 There are 74 parakeets and 38 parrots at the pet store.
How many more parakeets than parrots are there?

4 A baker made 300 donuts.
After selling some of them,
she had 85 donuts left.
How many donuts did she sell?

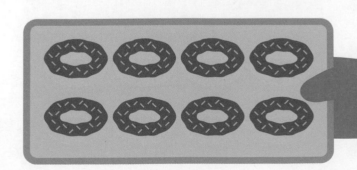

Exercise 11 • page 83

1 Find the value.

(a) 397 + 8

(b) 789 – 2

(c) 403 + 69

(d) 89 – 22

(e) 603 + 91

(f) 276 – 30

(g) 213 + 224

(h) 375 – 12

(i) 856 – 182

(j) 207 – 139

2 Xavier is adopting a cat and a dog from a shelter.
It costs $307 to adopt the dog.
It costs $96 to adopt the cat.

(a) How much less does it cost to adopt the cat than the dog?

(b) How much money does he need to adopt both pets?

3 Rea made 488 buttons to sell.
Sasha made 135 fewer buttons than Rea.

(a) How many buttons did Sasha make?

(b) How many buttons did they make altogether?

4 True or false?

(a) $64 + 72 = 74 + 62$

(b) $78 - 52 = 87 - 25$

(c) $256 + 37 < 275 + 73$

(d) $183 - 61 < 183 + 8$

(e) $100 + 54 = 200 - 45$

(f) $600 - 102 > 498 + 102$

(g) $267 + 499 < 456 + 389$

(h) $914 - 186 > 257 + 176$

5 There are 800 beads in a jar.
410 of them are black and the rest are orange.

(a) Are there more black beads or orange beads?

(b) How many more are there of one color than the other?

6 William saved $490.
He saved $60 more than Noah.
How much did they both save?

Exercise 12 • page 85

3-12 Practice C

Chapter 4

Length

Think

Sofia and Emma measured the lengths of different ribbons.

Emma's blue ribbon is about 7 paper clips long.

10 is greater than 7, but Emma's ribbon looks longer.

Sofia's orange ribbon is 10 blocks long.
What is a better way to compare the lengths of the ribbons?

Learn

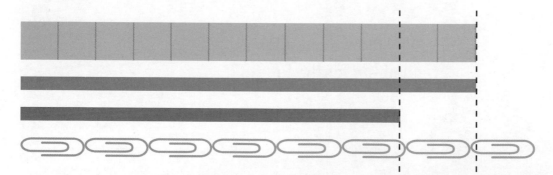

To compare lengths, we need to use the same unit to measure.

The blue ribbon is [   ] blocks longer than the orange ribbon.

The orange ribbon is about [   ] paper clip shorter than the blue ribbon.

The **centimeter** is a unit of length.
We write **cm** for centimeter.

 This cube is 1 centimeter long on each edge.

1 cm

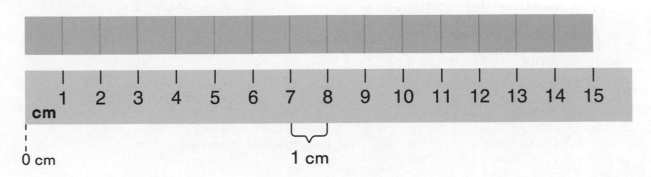

cm

0 cm   1 cm

The distance from one tick mark to the next on this ruler is 1 centimeter.

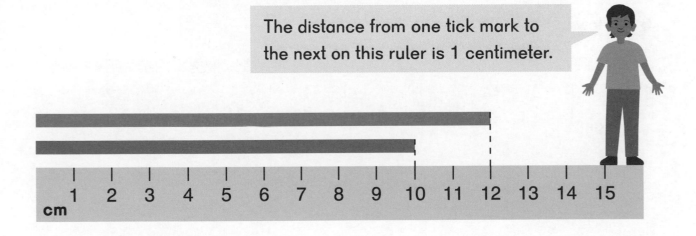

cm

The orange ribbon is 10 cm long.

The blue ribbon is ___ cm long.

The blue ribbon is ___ cm longer than the orange ribbon.

The total length of the blue and orange ribbons is ___ cm.

Do

1 What is the length of this red ribbon?

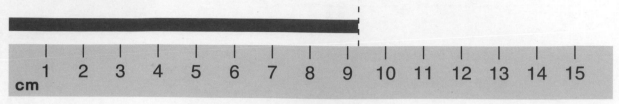

The length of the red ribbon is between ☐ cm and ☐ cm.
Its length is closer to ☐ cm.
It is about ☐ cm long.

If the measurement is between two units,
we can use the word **about** and the closest number.

2 How long is this pipe cleaner?

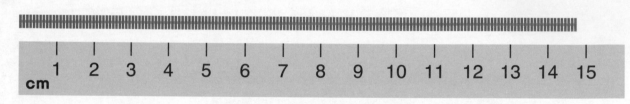

The piper cleaner is about ☐ cm long.

3 Measure the **length** and **width** of a piece of paper with a centimeter ruler.

width

length

4 (a) How long is the crayon?

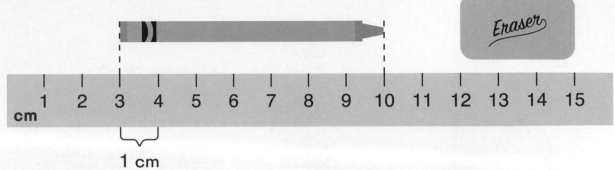

1 cm

We can line up the object with any mark. There are 7 centimeter units between 3 cm and 10 cm.

The crayon is ☐ cm long.

(b) The eraser is ☐ cm long.

(c) How much longer is the crayon than the eraser?

☐ ◯ ☐ = ☐

The crayon is ☐ cm longer than the eraser.

5 Use a centimeter ruler to draw a line that has a length of ...

(a) 9 cm

(b) 15 cm

How much longer is the 15-cm line than the 9-cm line?

Think

I know my finger is about 5 cm long. So I think her foot is about 15 cm long.

How many centimeters long do you think your foot is?

Measure your foot.

How close was your guess?

Learn

When we make an **estimate**, we try to find a number that is close to the right number without measuring.

If we know how long something familiar is, we can use that to help us estimate a length.

Do

1 Measure the distance from the tip of your thumb to the tip of your fifth finger in centimeters. Is it more or less than 10 cm?

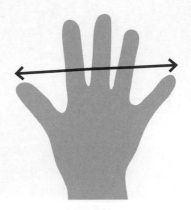

2 Draw a line that you think is about 10 cm long. Then measure it.

3 Look for something that you think has a length of about each of the following. Then use a centimeter ruler to check.

(a) 1 cm   (b) 10 cm

(c) 15 cm   (d) 30 cm

4 Estimate the length of the following objects in centimeters. Then use a centimeter ruler to find their lengths.

(a) The length of your pencil.

(b) The length of your math textbook.

(c) The width of your math textbook.

width

length

Exercise 2 • page 91

Think

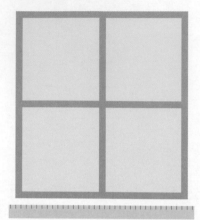

Use a meter stick to find things that are about 1 meter long.

Learn

We use **meters** to measure longer things.
We write **m** for meters.
100 cm = 1 m

This cabinet is a little more than 1 m high.

Do

1 About how many meters is the length and width of this living room?

1 m {

(a) The length of this living room is between ☐ m and ☐ m.
It is a little more than ☐ m long.

(b) The width of the living room is between ☐ m and ☐ m.
It is a little less than ☐ m long.

2 Measure a ribbon with a meter stick.
Mark every meter on the ribbon.
Cut the ribbon at 3 m.
Use the ribbon to find something around you that has a length of ...

(a) A little less than 2 m.

(b) Between 2 m and 3 m.

Exercise 3 • page 93

Think

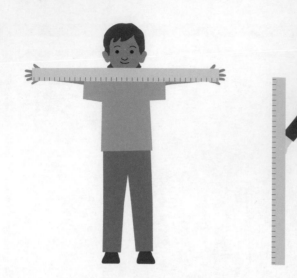

Is your arm span longer than a meter?

Estimate where on your body is 1 m from the floor.

Then check the height by measuring with a meter stick.

Estimate the length and width of your classroom.

Then measure.

Learn

We can use the length of something we are familiar with to estimate the length of other things.

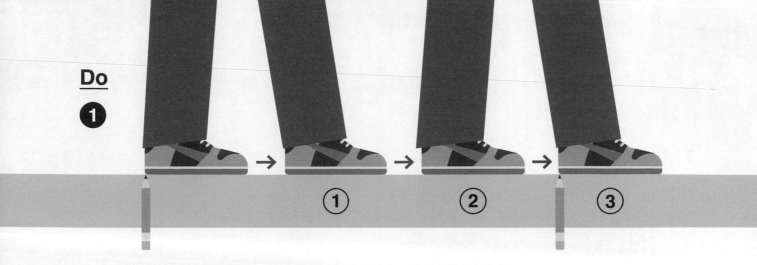

Do

1

Mark where your heel is.

Take 3 normal paces.

Mark where your heel is.

Is 3 paces more or less than 1 m?

Measure to find out.

2 Estimate the length of the following objects in meters.
Then find their lengths in meters.

(a) The length of the white board.

(b) The height of a doorway.

(c) The width of a window frame.

3 Which is a reasonable estimate for each?

A	The height of a doorway	2 m	6 m	10 m
B	The height of a 3-story building	4 m	12 m	100 m
C	The length of a telephone pole	3 m	8 m	20 m
D	The length of a soccer field	20 m	100 m	600 m

Exercise 4 · page 95

Think

The length of this paper is between 11 and 12 inches.

Use a ruler to find things that are less than 12 inches.
About how many inches is each object you measured?

Learn

In the United States, we also use the **inch** as a unit of length.
We write **in** for inch or inches.

The square face of this tile
is 1 inch long on each edge.

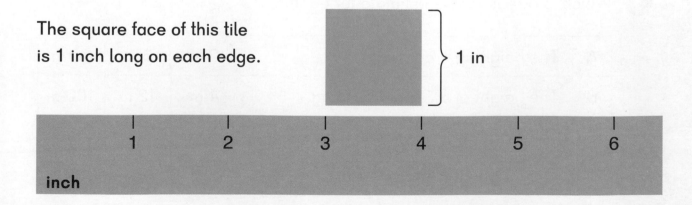

Do

1

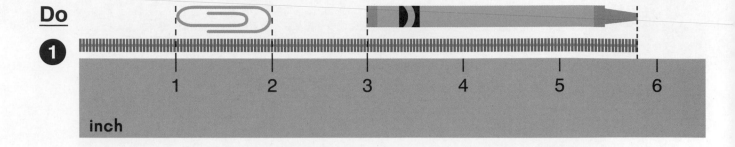

inch

(a) How long is the pipe cleaner?

(b) How long is the paper clip?

(c) How long is the crayon?

(d) How much longer is the pipe cleaner than the crayon?

2 Use an inch ruler to measure ...

(a) The length of your pencil.

(b) The length of your math textbook.

(c) The width of your math textbook.

3 Which of your finger joints is about 1 inch long?

4 Use an inch ruler to draw a line that has a length of:

(a) 11 in   (b) 4 in

How much shorter is the 4-inch line than the 11-inch line?

Exercise 5 • page 97

Think

Look at a ruler that has both centimeters and inches.

Which tick marks are for inches?

Which tick marks are for centimeters?

Where is 0 on the ruler?

Measure some things with the ruler in both centimeters and inches.

Learn

Compare 1 inch with 1 centimeter.
Which is longer?

<u>Do</u>

1 Which one shows the length of the rectangle
being measured in inches correctly?

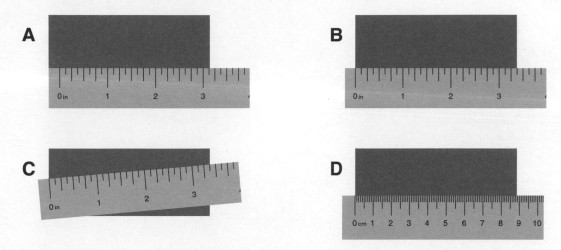

A

B

C

D

2 Estimate the length of the following objects in inches and in centimeters.
Then use a ruler to find their lengths.

(a) The length of the palm of your hand.

(b) The length of your forearm.

(c) The length of the sole of your foot.

3 Which is a reasonable estimate for each?

A	The wingspan of a butterfly	4 in	12 in	20 in
B	The length of an envelope	2 in	8 in	25 in
C	The length of a hammer	3 in	8 in	20 in
D	The width of a cell phone	3 in	6 in	12 in

Exercise 6 · page 101

Think

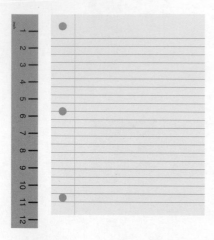

12 inches is the same as 1 foot.
This paper is almost 1 foot long.

Use a ruler to measure things around you in feet.
About how many feet is each object you measured?
Is your own foot longer or shorter than 1 foot?

Learn

In the United States, we also use the **foot** as a unit of length.
We write **ft** for foot or feet.
12 in = 1 ft

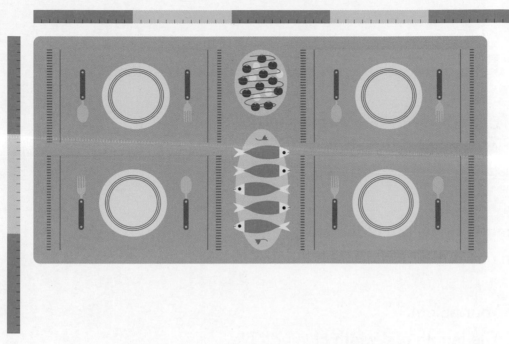

(a) The length of the table top is between ☐ ft and ☐ ft.
It is closer to ☐ ft.

(b) The width of the table top is between ☐ ft and ☐ ft.
It is closer to ☐ ft.

2 About how long and wide is this desk top?

1 ft

3 Find a part of your body that is about 1 ft long.

4 Use a straight edge to draw a line you think is 1 ft long.

(a) Use a ruler to see how close your line is to 1 ft.
(b) About how long is it in inches?

5 Estimate the length of the following objects in feet.
Then use a ruler to find their lengths in feet.

(a) Your height.
(b) The length and width of your table.
(c) The length and width of the white board.

6 Mark where your heel is.
Take 5 normal paces.
Mark where your heel is each time.
Measure the distance from one heel mark to the next in feet.
About how many feet is 1 pace?

7 Use string.

We call a length that is
3 ft long 1 **yard**.

(a) Estimate a 3-ft length of string.
(b) Check how close your estimate is with a ruler.
(c) Cut a 3-ft length of string.
(d) About how long is it in inches?
(e) About how long is it in meters?
(f) About how long is it in centimeters?

Exercise 7 • page 103

4-7 Feet

1 (a) 18 in + 16 in = ⬚ in

(b) 45 cm + 37 cm = ⬚ cm

(c) 34 in − 19 in = ⬚ in

(d) 83 cm − 57 cm = ⬚ cm

(e) 428 cm + 372 cm = ⬚ cm

(f) 305 m − 187 m = ⬚ m

(g) 177 ft + 208 ft = ⬚ ft

(h) 300 ft − 118 ft = ⬚ ft

2 Which of the following are good estimates for the height of a door?

7 ft   3 ft   7 m   3 m

3 Which of the following are good estimates for the length of a cell phone?

16 in   6 in   16 cm   6 cm

4 Simone walks all the way around the park.
How far does she walk?

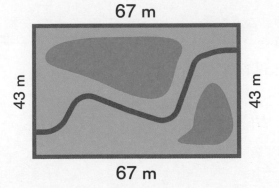

67 m

43 m

43 m

67 m

5 The distance between Maria's house and the school is 470 m.
Maria walked to school and then back home.
How far did she walk?

6 Phillip had 50 ft of string.
He used 32 ft of it to wrap packages.
How many feet of string did he have left?

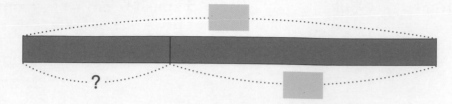

7 A red ribbon is 304 cm long.
A blue ribbon is 67 cm shorter than the red ribbon.

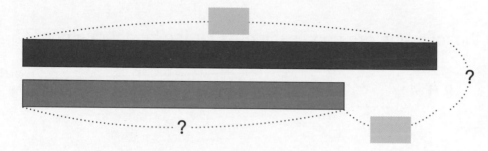

(a) How long is the blue ribbon?

(b) How long are both ribbons together?

8 Fuyu is 131 cm tall.
Mikhail is 124 cm tall.

(a) Who is taller?   (b) How much taller?

9 Imani ran 100 m.
Kalama ran 100 ft.
Who ran farther?

Exercise 8 • page 105

4-8 Practice

Chapter 5

Weight

Think

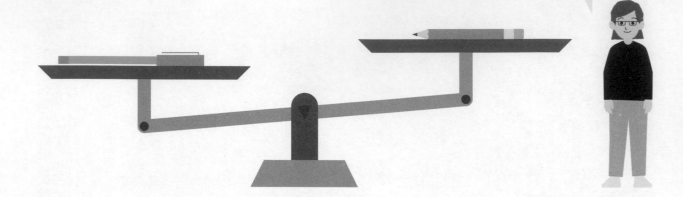

We can see which one weighs more, but how can we tell how much more?

Which weighs more, the pen or the pencil?

Learn

We weigh things to find out how heavy they are. The heavier they are, the more **mass** they have.

The **gram** is a unit of mass.
We write **g** for grams.

 ⟵ A paper clip weighs about 1 g.

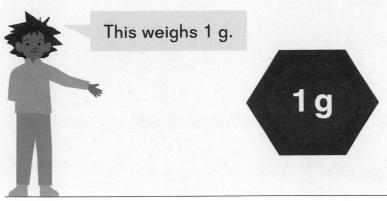

This weighs 1 g.

1 g

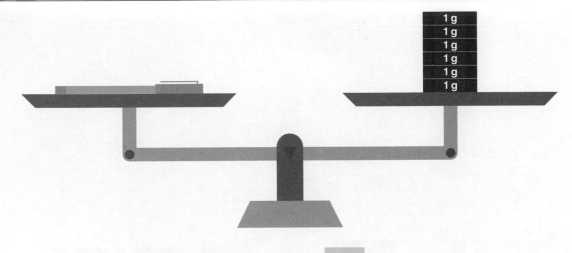

The pen weighs ⬜ g.

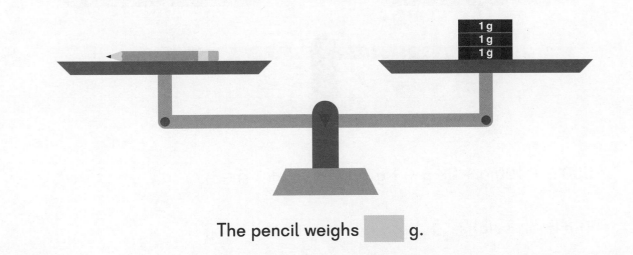

The pencil weighs ⬜ g.

6 g – 3 g = ⬜ g

The pen is ⬜ g heavier than the pencil.

Do

1

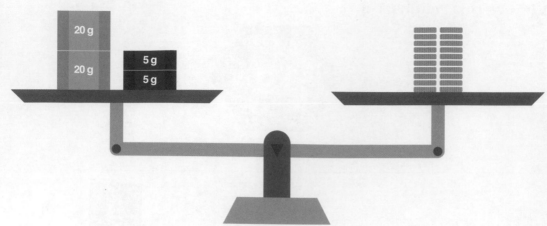

20 g + 20 g + 5 g + 5 g = [ ] g

The 20 pennies weigh [ ] g.

2

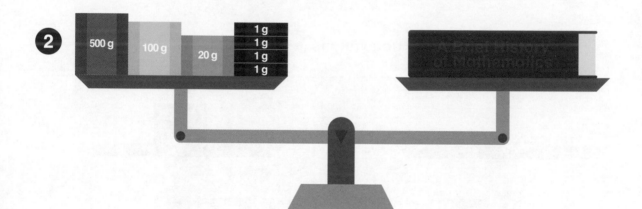

500 g + 100 g + 20 g + 1 g + 1 g + 1 g + 1 g = [ ] g

The book weighs [ ] g.

3 Use the balance scale to weigh your math textbook in grams.

4 Look for objects that you think weigh the following.
Then use the balance scale to check.

(a) 10 g   (b) 50 g

(c) 100 g   (d) 500 g

5 The mass of the stapler is 230 g.
What is the mass of the tape dispenser?

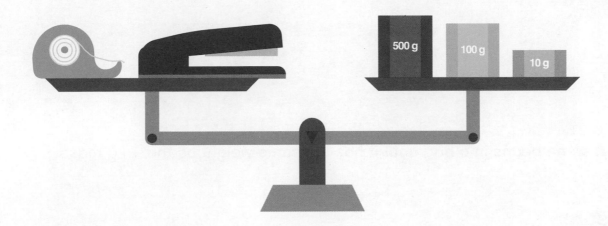

 ⬜ ◯ ⬜ = ⬜

The tape dispenser weighs ⬜ g.

6 A thermos filled with water weighs 902 g.
The empty thermos weighs 340 g.
How much does the water weigh?

⬜ ◯ ⬜ = ⬜

The water weighs ⬜ g.

Exercise 1 · page 109

Think

This is a 1 kilogram weight.

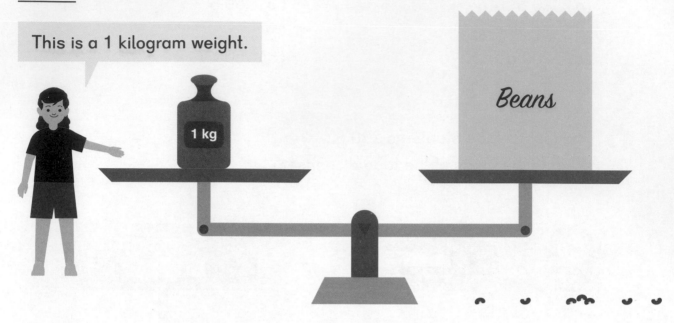

Put some beans in a bag until it has the same weight as the 1 kg mass.

Learn

We use **kilograms** to weigh heavier objects. We write **kg** for kilograms.

I weigh about 23 kg.

This bottle of sparkling water weighs 1 kg.

Do

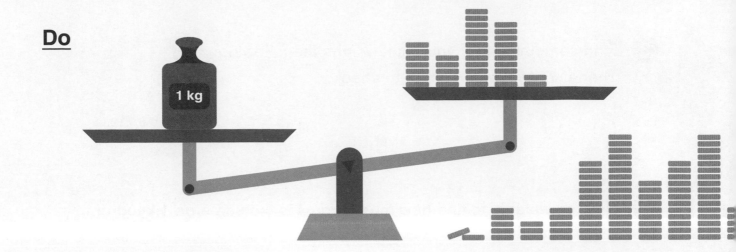

1. Use a balance scale to find out ...

   (a) How many pennies weigh about 1 kg.

   (b) How many crayons weigh about 1 kg.

   (c) How many notebooks weigh about 1 kg.

2. Look for objects that you think weigh ...

   (a) Less than 1 kg.

   (b) About 1 kg.

   (c) More than 1 kg.

   Then use the balance scale to check.

3 Find something that you think weighs the following.
Then use a weighing scale to check.

(a) 3 kg

(b) 5 kg

4 Use the weights to find how many grams is the same as 1 kilogram.

1 kg = ⬜ g

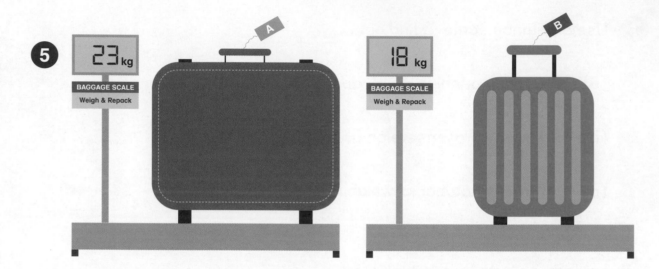

5

(a) How much more does suitcase A weigh than suitcase B?

⬜ ◯ ⬜ = ⬜

Suitcase A weighs ⬜ kg more than suitcase B.

(b) How much do they weigh altogether?

⬜ ◯ ⬜ = ⬜

They weigh ⬜ kg altogether.

Exercise 2 • page 113

5-2 Kilograms

Think

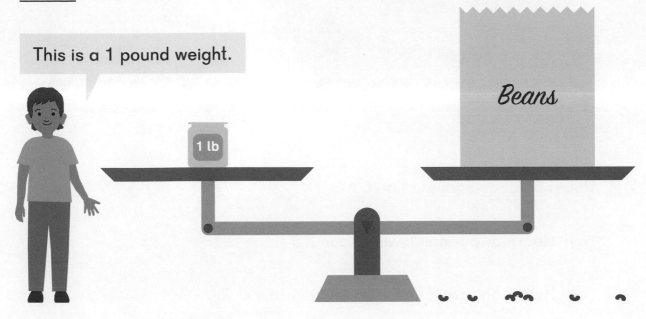

This is a 1 pound weight.

1 lb

Beans

Put some beans in a bag until it weighs 1 lb.

Learn

> In the United States, we also weigh things in **pounds**.
> We write **lb** for pounds.

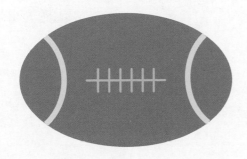

A football weighs about 1 lb.

KIDNEY
BEANS

A can of beans weighs almost 1 lb.

<u>Do</u>

1 Use a balance scale to find out ...

   (a) How many pennies weigh about 1 lb.

   (b) How many crayons weigh about 1 lb.

   (c) How many notebooks weigh about 1 lb.

2 Look for objects that you think weigh ...

   (a) Less than 1 lb.

   (b) About 1 lb.

   (c) More than 1 lb.

Then use the balance scale to check.

3 Find out which is heavier, 1 lb or 1 kg.

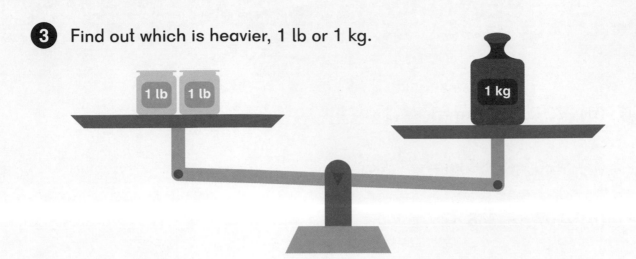

4 A Golden Retriever dog weighs 79 lb.
A Mastiff dog weighs 85 lb.

(a) Which dog weighs more, the Golden Retriever or the Mastiff?

(b) How much more?

(c) How much do they weigh together?

5 The potatoes weigh 11 lb.
The apples weigh 3 lb.
How much do the onions weigh?

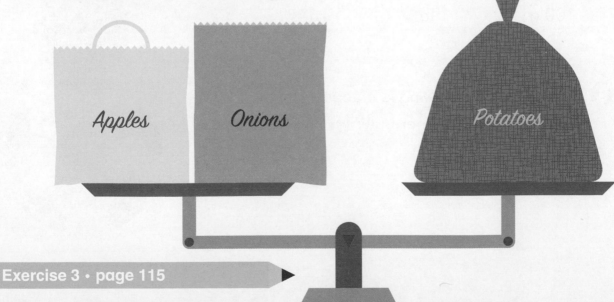

Exercise 3 • page 115

1 (a) 72 kg + 84 kg = ▢ kg

(b) 420 kg − 88 kg = ▢ kg

(c) 407 g + 348 g = ▢ g

(d) 306 lb − 192 lb = ▢ lb

(e) 38 lb + ▢ lb = 63 lb

(f) ▢ lb + 152 lb = 271 lb

(g) 125 kg − ▢ kg = 38 kg

(h) ▢ kg − 43 kg = 48 kg

2 Which of the following is a good estimate for the weight of an apple?

100 g   1 lb   1 kg

3 Which of the following is a good estimate for the weight of a soccer ball?

100 g   1 lb   1 kg

4 A baker used 280 kg of flour to bake biscuits
and 275 kg of flour to bake muffins.
How much flour did he use altogether?

5 Anthony weighs 92 lb.
He weighs 74 lb less than his father.
How much does his father weigh?

6 A bag of sugar weighs 125 g.
More sugar was added.
The bag of sugar now weighs 260 g.
How much sugar was added?

7 Jasmine's suitcase weighs 38 lb.
Laila's suitcase weighs 45 lb.

(a)   How much heavier is Laila's suitcase than Jasmine's suitcase?

(b)   What is the total weight of both suitcases?

8 A melon weighs 420 g.
A mango weighs 140 g less than the melon.

(a)   How much does the mango weigh?

(b)   What is the total weight of both fruits together?

Exercise 4 • page 117

1 What sign, > or <, goes in the ◯?

(a) 397 ◯ 401

(b) 659 ◯ 673

(c) 824 ◯ 821

(d) 490 ◯ 94

(e) 700 + 20 + 5 ◯ 70 + 200 + 5

(f) 500 + 5 ◯ 50 + 500

2 Arrange the numbers in order from least to greatest.

| 507 | 452 | 512 | 600 | 425 |

3 What number is:

(a) 3 more than 287

(b) 20 less than 310

(c) 2 less than 601

(d) 10 more than 590

4 Find the value.

(a) 81 − 43

(b) 70 − 57

(c) 460 + 88

(d) 460 − 88

(e) 506 − 235

(f) 506 + 235

(g) 615 + 276

(h) 615 − 276

(i) 285 + 78

(j) 200 − 102

(k) 700 − 312

(l) 498 + 289

5

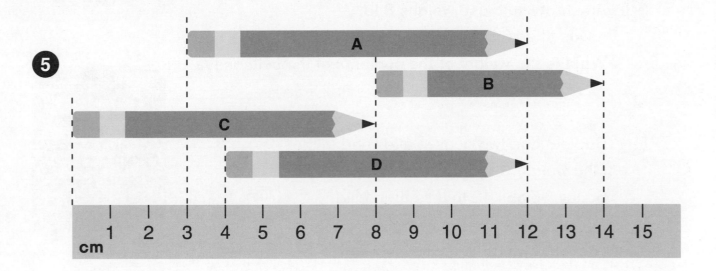

(a) How long is Pencil A?

(b) Which pencil is the longest?

(c) Which 2 pencils are the same length?

(d) Pencil B is 3 cm shorter than which pencil?

6 Abigail cut a thread into 2 pieces.
One piece is 141 cm long and the other is 59 cm long.
How long was the thread at first?

7 Aidan used 475 cm of ribbon to decorate presents.
280 cm was blue ribbon and the rest was red ribbon.
How much red ribbon did he use?

8 A soccer ball weighs 430 g and a tennis ball weighs 65 g.
How much less does the tennis ball weigh than the soccer ball?

9 An empty suitcase weighs 8 lb.
It was filled for a trip and now weighs 43 lb.
What is the weight of the contents of the suitcase?

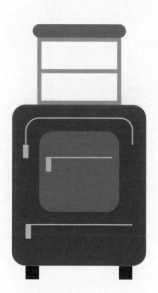

10 Daniela has two dogs, Bailey and Mila.
Mila weighs 110 lb.
Bailey weighs 42 lb less than Mila.

(a) How much does Bailey weigh?

(b) How much do the two dogs weigh together?

11 Colton ran 400 m.
His sister ran 50 m more than he did.
What is the total distance they ran altogether?

Exercise 5 • page 119

Chapter 6

Multiplication and Division

Think

How many hair ties are in the packages altogether?

Learn

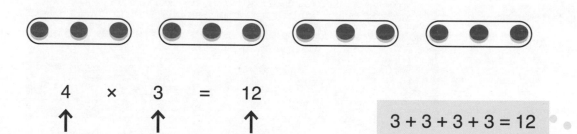

4 × 3 = 12

↑   ↑   ↑

number   number in   total
of groups   each group

3 + 3 + 3 + 3 = 12

There are ▢ hair ties altogether.

We multiply to find the total
when we have equal groups.
This is called **multiplication**.
× is read as **times**.

4 groups of 3 is 12.
4 times 3 equals 12.

Do

1 (a) There are 6 groups of 2 counters.

$2 + 2 + 2 + 2 + 2 + 2 =$

$6 \times 2 =$

(b) There are 4 groups of 5 blueberries.

$5 + 5 + 5 + 5 =$

$4 \times 5 =$

2

There are 5 paper clips in each box.
There are 6 boxes.
How many paper clips are there altogether?

$6 \times 5 =$

There are [ ] paper clips altogether.

3 There are 4 branches.
Each branch has 4 birds.
How many birds are there in all?

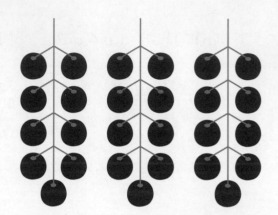

$4 \times 4 =$ [ ]

There are [ ] birds in all.

4 There are 3 bunches of grapes.
There are 9 grapes in each bunch.
How many grapes are there altogether?

[ ] × [ ] = [ ]

There are [ ] grapes altogether.

5 There are 10 oil pastels in each box.
There are 4 boxes.
How many oil pastels are there in all?

[ ] × [ ] = [ ]

There are [ ] oil pastels in all.

Oil Pastels SET OF 10
Oil Pastels SET OF 10
Oil Pastels SET OF 10
Oil Pastels SET OF 10

Exercise 1 • page 123

Think

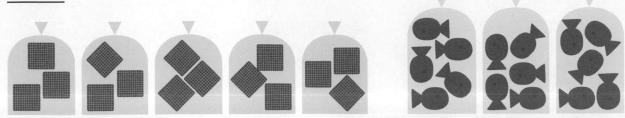

Mei has 5 bags with 3 square crackers in each bag.
Sofia has 3 bags with 5 fish crackers in each bag.
How many crackers does each friend have?

Learn

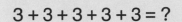

3 + 3 + 3 + 3 + 3 = ?

$5 \times 3 =$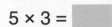

Mei has  square crackers.

5 + 5 + 5 = ?

$3 \times 5 =$

Sofia has [ ] fish crackers.

5 groups of 3 and 3 groups of 5 have the same total.

Do

1

$2 \times 4 = 4 \times 2$

$2 + 2 + 2 + 2 = \boxed{}$ | $4 \times 2 = \boxed{}$

$4 + 4 = \boxed{}$ | $2 \times 4 = \boxed{}$

2

$8 \times 3 = \boxed{}$

$3 \times 8 = \boxed{}$

$8 \times 3 = 3 \times \boxed{}$

3

$4 \times 6 = \boxed{}$

$6 \times 4 = \boxed{}$

$4 \times 6 = \boxed{} \times 4$

4

$8 \times 2 = \boxed{}$

$2 \times 8 = \boxed{}$

Exercise 2 • page 127

1 (a) $6 + 6 =$ ☐

$2 \times 6 =$ ☐

(b) $2 + 2 =$ ☐

$2 \times 2 =$ ☐

(c) $5 + 5 + 5 + 5 =$ ☐

$4 \times 5 =$ ☐

(d) $3 + 3 + 3 =$ ☐

$3 \times 3 =$ ☐

(e) $7 \times 2 =$ ☐

$2 \times 7 =$ ☐

$7 \times 2 =$ ☐ $\times 7$

(f) $3 \times 4 =$ ☐

$4 \times 3 =$ ☐

$3 \times 4 = 4 \times$ ☐

(g) $8 \times 6 = 6 \times$ ☐

(h) $9 \times 7 =$ ☐ $\times 9$

2

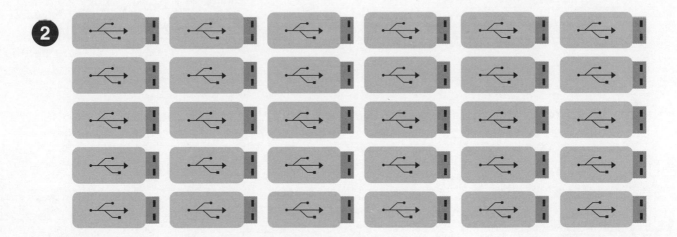

☐ $\times 5 =$ ☐

$6 \times$ ☐ $=$ ☐

3 Write two different multiplication equations to find the number of clownfish.

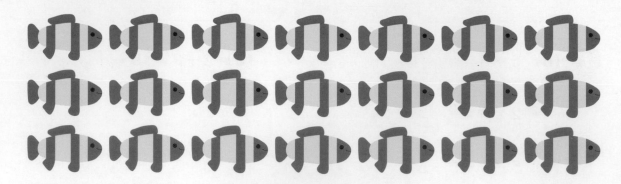

4 Write two different multiplication equations to find the number of bowls.

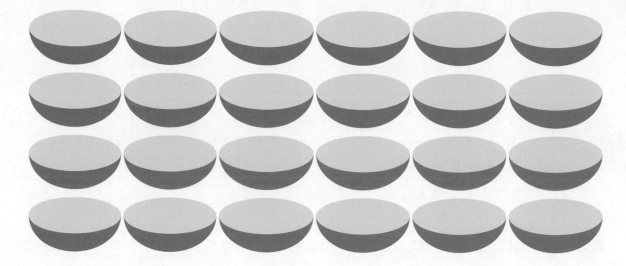

5 Write two different multiplication equations to find the number of beans.

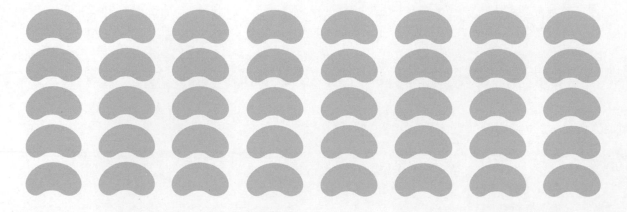

6 Write a multiplication equation and find the answer.

(a) There are 3 cotton swabs in 1 can.
How many cotton swabs
are in 3 cans?

(b) There are 10 boxes of oranges.
There are 4 oranges in each box.
How many oranges are there in all?

(c) One brick weighs 5 kg.
How much do 2 bricks weigh?

(d) One milkshake costs $3.
How much will 10 milkshakes cost?

(e) June walked 2 miles every day for 7 days.
What is the total distance that she walked?

Exercise 3 • page 131

Think

3 children share 12 snickerdoodle cookies equally.
How many will each child get?

Learn

If each child gets 1 snickerdoodle cookie ...

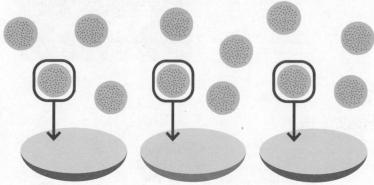

If each child gets 2 snickerdoodle cookies ...

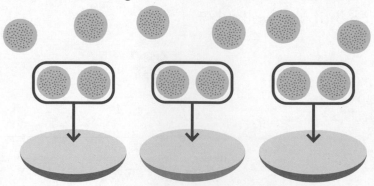

There are still snickerdoodle cookies to share.

If each child gets 3 snickerdoodle cookies ...

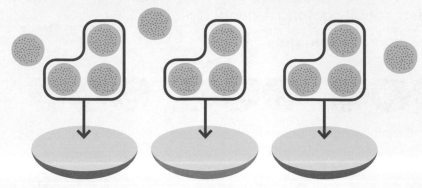

If each child gets 4 snickerdoodle cookies ...

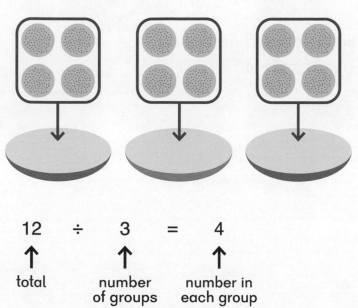

There are no more snickerdoodle cookies to share.

$$12 \div 3 = 4$$

↑ total   ↑ number of groups   ↑ number in each group

Each child gets ☐ snickerdoodle cookies.

I know the total and the number of groups.

We divide to find the number in each equal group.
This is called **division**.
÷ is read as **divided by**.

<u>Do</u>

1 Emma shares 12 cherries equally among 4 people.
How many cherries does each person get?

12 ÷ 4 = ▢

Each person gets ▢ cherries.

2 Dion divides 20 cashews equally among 5 people.
How many cashews does each person get?

20 ÷ 5 = ▢

Each person gets ▢ cashews.

3 Make 3 towers using 27 cubes.
How many cubes are in each tower?

27 ÷ ▢ = ▢

There are ▢
cubes in each tower.

4 There are 16 hot peppers.
Put them equally onto 4 plates.
How many hot peppers are on each plate?

	÷		=	

There are ⬜ hot peppers on each plate.

5 Mei divides 30 flowers equally between 6 vases.
How many will be in each vase?

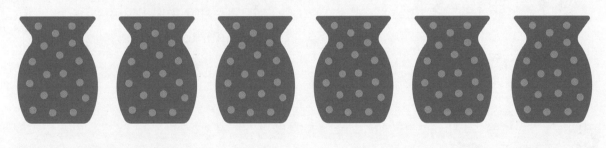

	÷		=	

There will be ⬜ flowers in each vase.

6 A pet shop owner puts 18 goldfish equally in 3 bowls.
How many goldfish are in each bowl?

	÷		=	

There are ⬜ goldfish in each bowl.

Exercise 4 • page 135

Think

There are 12 lychees.

If 3 lychees are put in each bowl, how many bowls are needed?

Learn

If 1 bowl gets 3 lychees …

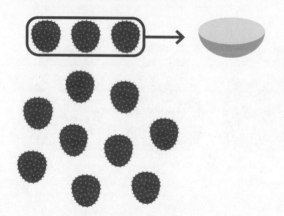

There are still lychees
to put into groups.

If 2 bowls get 3 lychees …

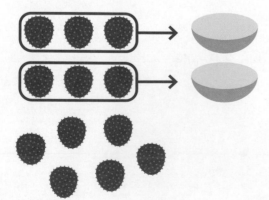

If 3 bowls get 3 lychees ...

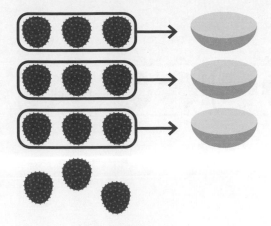

If 4 bowls get 3 lychees ...

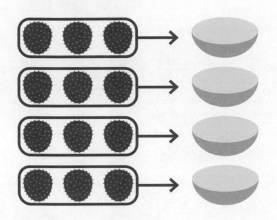

There are no more lychees to put into groups.

$$12 \div 3 = 4$$

↑ total

↑ number in each group

↑ number of groups

I know the total and the number in each group.

_____ bowls are needed.

We also divide to find the number of equal groups.

Do

1 Alex divides 12 buttons into groups of 4.
How many groups are there?

$12 \div 4 = $ ▢

There are ▢ groups.

2 Sofia divides 15 tortilla chips into groups of 5.
How many groups are there?

$15 \div 5 = $ ▢

There are ▢ groups.

3 Each vine has 3 cranberries.
There are 21 cranberries.
How many vines are there?

$21 \div $ ▢ $ = $ ▢

There are ▢ vines.

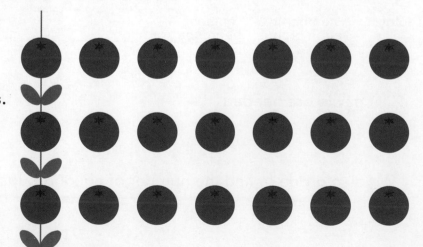

4 There are 25 crayons.

Put 5 crayons in each box.

How many boxes of 5 crayons are there?

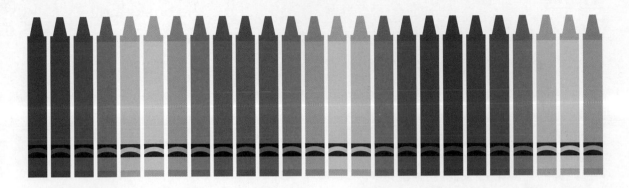

[ ] ÷ [ ] = [ ]

There are [ ] boxes.

5 Dion has 24 craft sticks.

He uses 4 sticks for each frame.

How many frames does he make?

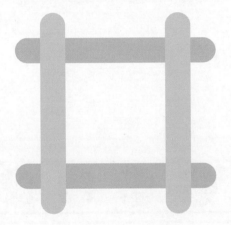

[ ] ÷ [ ] = [ ]

He makes [ ] frames.

6 A pet shop owner wants to put 5 hamsters in each cage.

There are 20 hamsters.

How many cages does she need?

[ ] ÷ [ ] = [ ]

She needs [ ] cages.

Exercise 5 • page 139

Think

Write different multiplication and division equations for the 15 basketballs.

The basketballs are arranged in an **array**.

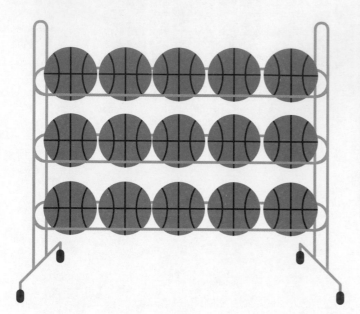

I can make equal groups two different ways, but the total is still the same.

Learn

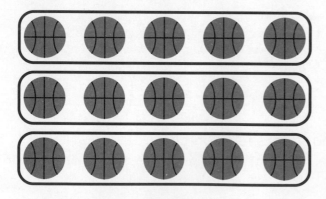

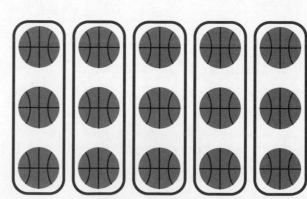

3 × 5 =

15 ÷ 3 =

5 × 3 =

15 ÷ 5 =

Do

1 Complete the multiplication and division equations for the array of watermelon slices.

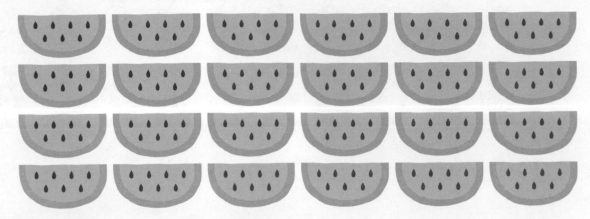

$4 \times 6 =$ [ ] $6 \times 4 =$ [ ]

$24 \div 4 =$ [ ] $24 \div 6 =$ [ ]

2 Write four related multiplication and division equations for each array.

(a)

$2 \times$ [ ] $=$ [ ] $4 \times$ [ ] $=$ [ ]

[ ] $\div$ [ ] $=$ [ ]   [ ] $\div$ [ ] $=$ [ ]

(b)

$3 \times$ [ ] $=$ [ ] | $10 \times$ [ ] $=$ [ ]

[ ] $\div$ [ ] $=$ [ ] | [ ] $\div$ [ ] $=$ [ ]

(c)

$4 \times$ [ ] $=$ [ ]

[ ] $\div$ [ ] $=$ [ ]

Why can we write only 1 multiplication and 1 division equation for this array?

3 Draw an array of objects.
Write two multiplication and two division equations for your drawing.

Exercise 6 • page 143

6-6 Multiplication and Division

1 Write a multiplication or division equation for each story and find the answer.

(a) 5 children share 30 carrot pieces equally.
How many carrot pieces does each child get?

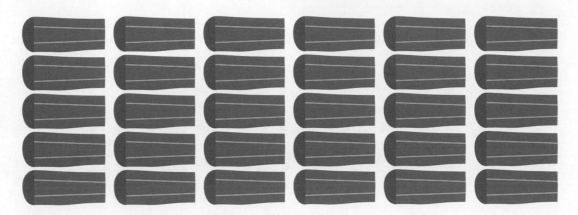

(b) There are 40 toothpicks.
Dana puts 8 toothpicks in each box.
How many boxes are there?

(c) There are 6 different colors of clay.
There are 3 pieces of clay of each color.
How many pieces of clay are there in all?

(d) There are 10 muffins.

2 muffins are put in each box.

How many boxes are there?

(e) There are 3 bags of potatoes.

Each bag has 10 potatoes.

How many potatoes are there altogether?

2 Write multiplication and division equations for the array of caterpillars.

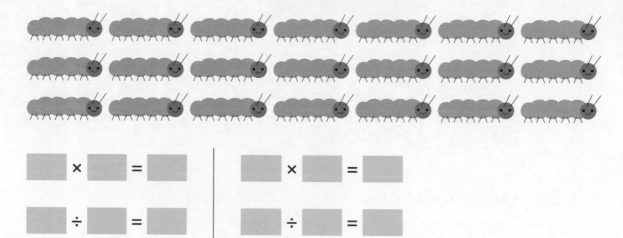

$$\boxed{} \times \boxed{} = \boxed{} \quad \Big| \quad \boxed{} \times \boxed{} = \boxed{}$$

$$\boxed{} \div \boxed{} = \boxed{} \quad \Big| \quad \boxed{} \div \boxed{} = \boxed{}$$

3 A piece of string is 12 cm long.

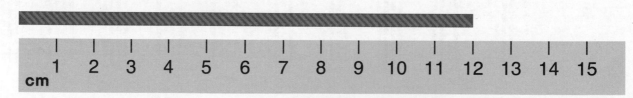

(a) If it is cut into 2 pieces of equal length,
how long will each piece be?

(b) If it is cut into pieces that are each 2 cm long,
how many pieces will there be?

Exercise 7 • page 145

6-7 Practice B

Chapter 7

Multiplication and Division of 2, 5, and 10

Think

There are 5 mangoes in each box.

$1 \times 5 = $ ▢

How many mangoes are there in ▢ boxes?

▢ × 5 = ?

How many mangoes are there in 2 boxes?

Find the number of mangoes if there are

3 , 4 , 5 , 6 , 7 , 8 , 9 , and 10 boxes.

How does the total number of mangoes change when the number of boxes increases by 1?

Learn

+1 ⎰ 1 × 5 = 5 ⎰ +5 ⎱ ●●●●● ⎱ +5
   ⎱ 2 × 5 = 10 ⎰   ●●●●●

3 × 5 = 15   ●●●●●   3 × 5 is ▢ more than 2 × 5.

4 × 5 = 20   ●●●●●   4 × 5 is ▢ less than 5 × 5.

5 × 5 = 25   ●●●●●

6 × 5 = 30   ●●●●●

7 × 5 = 35   ●●●●●   7 × 5 is 5 more than ▢ × 5.

8 × 5 = 40   ●●●●●

9 × 5 = 45   ●●●●●   9 × 5 is 5 less than ▢ × 5.

10 × 5 = 50   ●●●●●

What do you notice about the ones digit in the products?

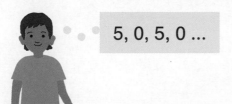 5, 0, 5, 0 ...

 50 is the **product** of 10 and 5.

Do

1 Count by 5s to 50.

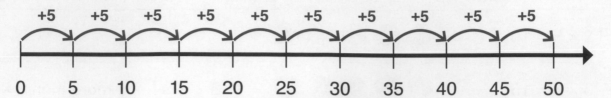

(a) 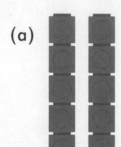 $2 \times 5 =$ [ ]

(b) 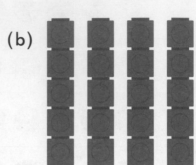 $4 \times 5 =$ [ ]

2 Use array dot cards to show the totals.

(a) $5 \times 5 =$ [ ]

(b) $10 \times 5 =$ [ ]

$6 \times 5 =$ [ ]

$9 \times 5 =$ [ ]

$7 \times 5 =$ [ ]

$8 \times 5 =$ [ ]

3 A nickel is worth 5 cents.

How many cents are 3 nickels worth?

[ ] × [ ] = [ ]

3 nickels are worth [ ] ¢.

Exercise 1 • page 149

7-1 The Multiplication Table of 5

Think

Alex has 4 bags with 5 strawberries in each bag.
Sofia has 5 bags with 4 strawberries in each bag.
How many strawberries does each friend have?

Learn

5 + 5 + 5 + 5 = [ ] | 4 × 5 = [ ]

Alex has [ ] strawberries altogether.

4 groups of 5 equals
5 groups of 4.
4 × 5 = 5 × 4

4 + 4 + 4 + 4 + 4 = [ ] | 5 × 4 = [ ]

Sofia has [ ] strawberries altogether.

Do

1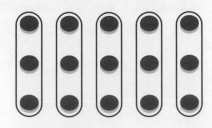

$5 + 5 + 5 =$ ▮ $3 + 3 + 3 + 3 + 3 =$ ▮

$3 \times 5 =$ ▮ $5 \times 3 =$ ▮

2 (a) $2 \times 5 =$ ▮ $\times 2$ (b) $7 \times$ ▮ $= 5 \times 7$

(c) $9 \times 5 = 5 \times$ ▮   (d) ▮ $\times 5 = 5 \times 6$

3 Mei made 5 kabobs for dinner.
Each kabob has 3 shrimp on it.
She gives 5 shrimp to Dion.

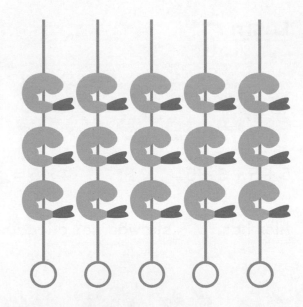

(a) How many shrimp
did she have at first?

$5 \times 3 =$ ▮

She had ▮ shrimp at first.

(b) How many shrimp does she have left?

▮ $- 5 =$ ▮

She has ▮ shrimp left.

4

How many of the 5-gram weights should we put on the right side to balance the left side?

5 What is the value of each?

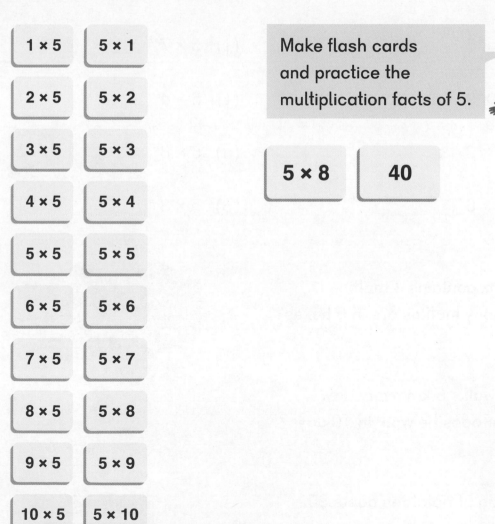

1 × 5	5 × 1
2 × 5	5 × 2
3 × 5	5 × 3
4 × 5	5 × 4
5 × 5	5 × 5
6 × 5	5 × 6
7 × 5	5 × 7
8 × 5	5 × 8
9 × 5	5 × 9
10 × 5	5 × 10

Make flash cards and practice the multiplication facts of 5.

5 × 8	40

Exercise 2 • page 151

1 Find the value.

(a) 2 × 5

(b) 4 × 5

(c) 8 × 5

(d) 3 × 5

(e) 6 × 5

(f) 9 × 5

(g) 5 × 5

(h) 10 × 5

(i) 7 × 5

(j) 5 × 7

(k) 5 × 9

(l) 5 × 4

(m) 5 × 2

(n) 5 × 8

(o) 5 × 6

(p) 5 × 3

2 One box contains 4 muffins.
How many muffins are in 5 boxes?

3 Henry walks 5 km every day.
How far does he walk in 10 days?

4 One bag of potatoes costs $5.
How much do 7 bags of potatoes cost?

5

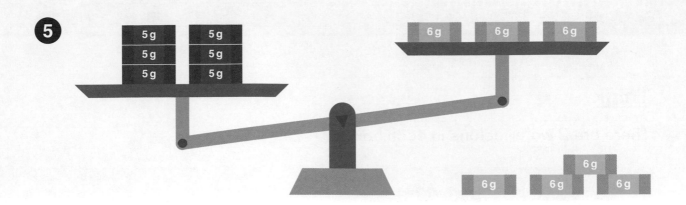

How many more of the 6-gram weights
do we need to make it balance?

6 Camila has 8 nickels.
She buys a ring that costs 15 cents.

(a) How much money did
she have at first?

$8 \times 5 =$ ⬜

She had ⬜ ¢ at first.

(b) How much money
does she have left?

⬜ $- 15 =$ ⬜

She has ⬜ ¢ left.

7 Dexter has 5 nickels.
He gets 2 more nickels.

(a) How many nickels does he have in all?

(b) How many cents does he have in all?

Exercise 3 • page 153

Lesson 4
The Multiplication Table of 2

④

Think

There are 2 watermelons in each box.

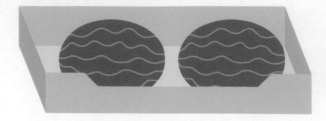

$1 \times 2 = $ ▢

How many watermelons are there in ▢ boxes?

▢ $\times 2 = ?$

How many watermelons are there in 2 boxes?

Find the number of watermelons if there are ...

3 , 4 , 5 , 6 , 7 , 8 , 9 , and 10 boxes.

How does the total number of watermelons change when the number of boxes increases by 1?

Learn

+1 ⎛ 1 × 2 = 2 ⎞ +2 ────── +2 ────────────────────
⎝ 2 × 2 = 4 ⎠

3 × 2 = 6   3 × 2 = 4 +

4 × 2 = 8

5 × 2 = 10

6 × 2 = 12   6 × 2 = 10 +

7 × 2 = 14

8 × 2 = 16

9 × 2 = 18   9 × 2 = 20 −

10 × 2 = 20

What do you notice about the ones digit in the products?

2, 4, 6, 8, 0 ...

Do

1 Count by 2s to 20.

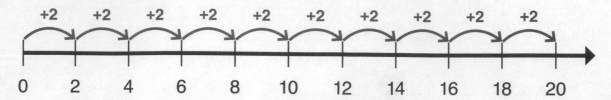

2 (a) $2 \times 2 =$ ☐ (b) $4 \times 2 =$ ☐

3 Use array dot cards to show the totals.

(a) $5 \times 2 =$ ☐   (b) $10 \times 2 =$ ☐

$6 \times 2 =$ ☐ $9 \times 2 =$ ☐

$7 \times 2 =$ ☐ $8 \times 2 =$ ☐

4 8 children each have $2.
How much money do they have altogether?

☐ × ☐ = ☐ | They have $ ☐ .

5 2 people are in each kayak.
How many people are in 7 kayaks?

Exercise 4 • page 157

Think

Emma has 6 boxes of 2 chocolates.

Mei has 2 boxes of 6 chocolates.

How many chocolates does each friend have?

Learn

$2 + 2 + 2 + 2 + 2 + 2 =$ ▢ | $6 \times 2 =$ ▢

Emma has ▢ chocolates altogether.

$6 + 6 =$ ▢ | $2 \times 6 =$ ▢

Mei has ▢ chocolates altogether.

> 6 groups of 2 equals
> 2 groups of 6.
> $6 \times 2 = 2 \times 6$

1

$2 + 2 + 2 + 2 =$ ☐ $4 + 4 =$ ☐

$4 \times 2 =$ ☐ $2 \times 4 =$ ☐

2 (a) $5 \times 2 =$ ☐ $\times 5$ (b) $9 \times$ ☐ $= 2 \times 9$

(c) $10 \times 2 = 2 \times$ ☐   (d) ☐ $\times 2 = 2 \times 8$

3

There are 2 rows of stickers.

There are 7 columns of stickers.

How many stickers are there altogether?

Write two different multiplication equations
to find the total number of stickers.

$2 \times$ ☐ $=$ ☐ $7 \times$ ☐ $=$ ☐

There are ☐ stickers altogether.

4 There are 8 packs of glue sticks.
There is one yellow and one red glue stick in each.
How many glue sticks are there altogether?
Write two different multiplication equations
to find the total number of glue sticks.

5 What is the value of each?

1 × 2	2 × 1
2 × 2	2 × 2
3 × 2	2 × 3
4 × 2	2 × 4
5 × 2	2 × 5
6 × 2	2 × 6
7 × 2	2 × 7
8 × 2	2 × 8
9 × 2	2 × 9
10 × 2	2 × 10

Make flash cards
and practice the
multiplication facts of 2.

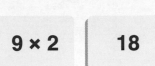
9 × 2   18

Exercise 5 • page 159

1 Find the value.

(a) 2 × 2

(b) 4 × 2

(c) 8 × 2

(d) 3 × 2

(e) 6 × 2

(f) 9 × 2

(g) 5 × 2

(h) 10 × 2

(i) 2 × 8

(j) 2 × 7

(k) 2 × 4

(l) 2 × 10

(m) 2 × 9

(n) 2 × 6

(o) 2 × 3

(p) 2 × 10

2 (a) How many wheels do 4 bicycles have?

(b) How many wheels do 8 bicycles have?

3 (a) How much does 3 lb of rice cost?

(b) How much does 6 lb of rice cost?

JASMINE
RICE

$5

1 lb

4

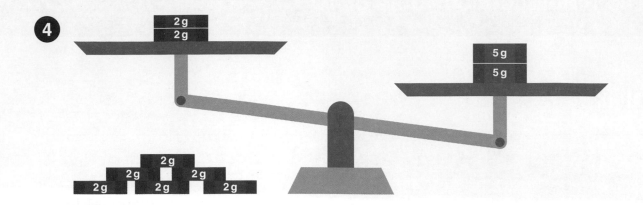

How many more of the 2-gram weights
do we need to make it balance?

5 A bag of flour weighs 2 kg.
How much do 6 bags of flour weigh?

6 A bag of peanuts weighs 2 lb.
Carlos had 10 bags of peanuts.
He gave 5 bags away.

(a) How many bags of peanuts does he have left?

(b) How much do the bags he has left weigh in all?

7 8 posts are in a straight line.
The distance from one post to the next is 2 m.
What is the total distance between the first and last post?

2 m

Exercise 6 • page 161

Think

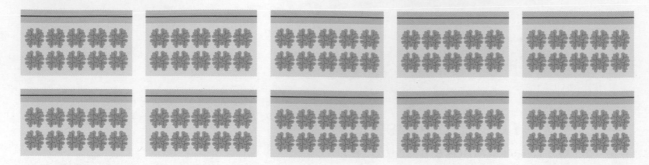

There are 10 walnuts in a bag.
How many walnuts are there in 10 bags?

$10 \times 10 =$ ⬜

How many walnuts are there in ⬜ bags?

⬜ $\times 10 = ?$

How many walnuts are there in │ 9 │ bags?

Find the number of walnuts if there are

│ 8 │, │ 7 │, │ 6 │, │ 5 │, │ 4 │, │ 3 │, │ 2 │, and │ 1 │ bags.

How does the total number
of walnuts change when the
number of bags decreases by 1?

Learn

1 × 10 = 10

2 × 10 = 20

3 × 10 = 30

4 × 10 = 40

5 × 10 = 50   5 × 10 = 40 +

6 × 10 = 60

7 × 10 = 70   7 × 10 = 80 −

8 × 10 = 80

−1 ⎛ 9 × 10 = 90 ⎞ −10   −10
   ⎝ 10 × 10 = 100 ⎠

What do you notice about
the ones digit in the products?

0, 0, 0, ...

Do

1 Count by 10s to 100.

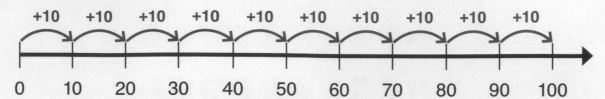

2

$7 \times 10 =$ ☐

3 Use array dot cards to show the totals.

(a) $5 \times 10 =$ ☐   (b) $10 \times 10 =$ ☐

  $6 \times 10 =$ ☐ $9 \times 10 =$ ☐

  $7 \times 10 =$ ☐ $8 \times 10 =$ ☐

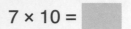

4 A dime is worth 10 cents.
How many cents are 7 dimes worth?

5

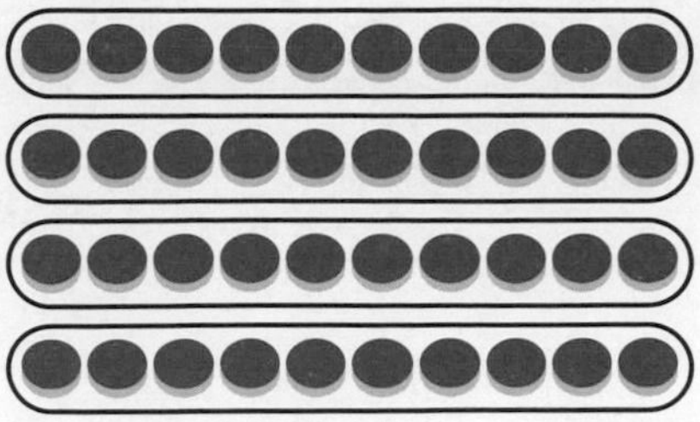

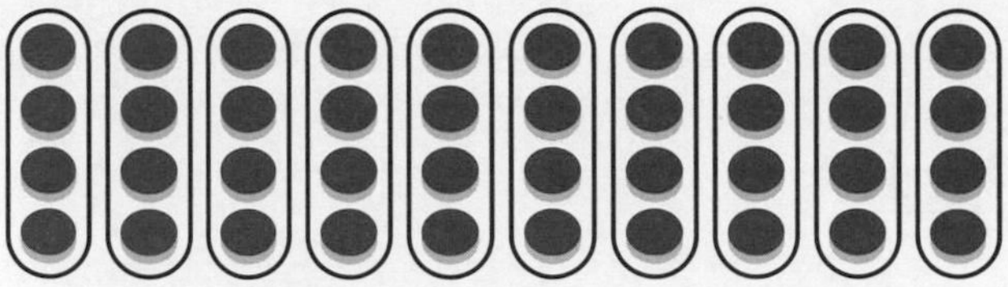

$10 + 10 + 10 + 10 =$ ___

$4 + 4 + 4 + 4 + 4 + 4 + 4 + 4 + 4 + 4 =$ ___

$4 \times 10 =$ ___

$10 \times 4 =$ ___

6 What is the value of each?

1 × 10	**10 × 1**
2 × 10	**10 × 2**
3 × 10	**10 × 3**
4 × 10	**10 × 4**
5 × 10	**10 × 5**
6 × 10	**10 × 6**
7 × 10	**10 × 7**
8 × 10	**10 × 8**
9 × 10	**10 × 9**
10 × 10	**10 × 10**

Make flash cards and practice the multiplication facts of 10.

10 × 7	**70**

Exercise 7 • page 165

Think

Mei and Alex want to share 12 strawberries equally.
How many will each get?

Learn

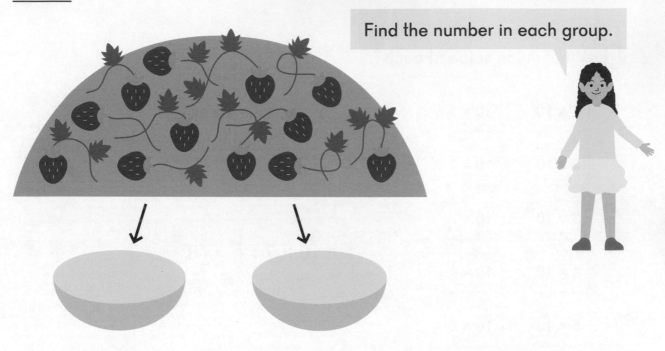

Find the number in each group.

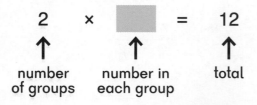

$$2 \quad \times \quad \boxed{} \quad = \quad 12$$

number of groups   number in each group   total

To divide by 2, we can use the multiplication facts of 2.
$2 \times 6 = 12$, so $12 \div 2 = 6$.

$$12 \quad \div \quad 2 \quad = \quad \boxed{}$$

total   number of groups   number in each group

Do

1 (a) Divide 16 counters into 2 equal groups.   2 × ? = 16

16 ÷ 2 =

There are [   ] counters in each group.

(b) Divide 16 counters into groups of 2.   ? × 2 = 16

16 ÷ 2 =

There are [   ] groups.

2 (a) [   ] × 2 = 8

8 ÷ 2 =

(b) 2 × [   ] = 14

14 ÷ 2 =

3 Find the value.

(a) $4 \div 2$

(b) $8 \div 2$

(c) $16 \div 2$

(d) $18 \div 2$

(e) $6 \div 2$

(f) $20 \div 2$

(g) $10 \div 2$

(h) $12 \div 2$

4 Alyssa has 18 ft of ribbon.
She cuts the ribbon into 2 equal-length pieces.
How long is each piece?

$18 \div 2 = \boxed{}$

Each piece is $\boxed{}$ ft long.

5 A baker has a 14-lb bag of flour and 2 bowls.
She wants to put the same amount of flour into each bowl.
How much flour will go in each bowl?

6 The baker has a 15-lb bag of sugar.
She wants to put all of the sugar into bowls.
Each bowl holds up to 2 lb of sugar.
How many bowls does she need if she wants
to use the fewest number of bowls?

Exercise 8 • page 167

7-8 Dividing by 2

Think

Emma has 15 celery sticks.
She puts 5 celery sticks on each plate.
How many plates does she need?

Learn

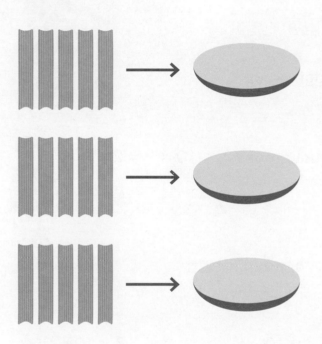

Find the number of groups.

⬛	×	5	=	15
↑		↑		↑
number of groups		number in each group		total

15 ÷ 5 = ⬛

15	÷	5	=	⬛
↑		↑		↑
total		number in each group		number of groups

To divide by 5, we can use the multiplication facts of 5.
3 × 5 = 15, so 15 ÷ 5 = 3.

Do

1 (a) Divide 20 counters into groups of 5.

$? \times 5 = 20$

$20 \div 5 =$ ▢

There are ▢ groups.

(b) Divide 20 counters into 5 equal groups.

$5 \times ? = 20$

$20 \div 5 =$ ▢

There are ▢ counters in each group.

(c) Divide 20 counters into 10 equal groups.

$10 \times ? = 20$

$20 \div 10 =$ ▢

There are ▢ counters in each group.

(d) Divide 20 counters into groups of 10.

$? \times 10 = 20$

$20 \div 10 =$ ▢

There are ▢ groups.

2 (a) 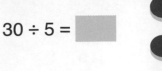 $\times\ 5 = 30$

$30 \div 5 =$

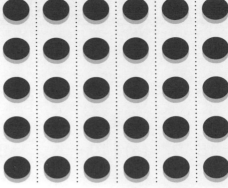

(b) $10 \times$ ⬜ $= 50$

$50 \div 10 =$

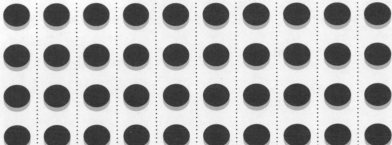

3 Find the value.

(a) $25 \div 5$

(b) $15 \div 5$

(c) $40 \div 5$

(d) $35 \div 5$

(e) $80 \div 10$

(f) $60 \div 10$

4 Adriana has 45 pencils.
She wants to put 5 pencils in each bundle.
How many rubber bands does she need?

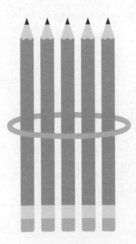

5 A carpenter has a board that is 20 ft long.
He saws it into 5 pieces of equal length.
How long is each piece?

Exercise 9 • page 171

① (a) ▢ × 2 = 12　　(b) 2 × ▢ = 14　　(c) 5 × ▢ = 25

(d) ▢ × 5 = 15　　(e) 10 × ▢ = 80　　(f) ▢ × 10 = 40

② Find the value.

(a) 10 ÷ 2　　　　(b) 50 ÷ 5　　　　(c) 14 ÷ 2

(d) 35 ÷ 5　　　　(e) 20 ÷ 2　　　　(f) 90 ÷ 10

(g) 40 ÷ 5　　　　(h) 100 ÷ 10　　　(i) 45 ÷ 5

(j) 18 ÷ 2　　　　(k) 60 ÷ 10　　　(l) 25 ÷ 5

③ A bowling ball weighs 10 lb.
How much do 2 similar bowling balls weigh?

④ Jaiden cut 16 ft of wire into 2 equal lengths.
How long is each piece?

⑤ Aurora has 82 beads.
She wants to make some necklaces.
Each necklace should have 10 beads.
How many necklaces can she make
that have 10 beads?

Think

Write an equation for each and find the answer.

Alex's allowance						
M	T	W	Th	F	S	S
						$5
						$5
						$5
						$5

Dion's allowance						
M	T	W	Th	F	S	S
						$?
						$?
						$?
						$?

Emma's savings						
M	T	W	Th	F	S	S
						$10
						$10
						$10
						$10

(a) Alex gets a $5 allowance each week.
How much money will he have in 6 weeks?

(b) After 5 weeks, Dion has $30 from his allowance.
He gets the same amount of money each week.
How much money does he get each week?

(c) Emma saves $10 each week.
How long will it take for her to save $50?

Learn

We need to find the total.
We have equal groups.

(a)

?

$5

$6 \times 5 =$

Alex will have $ ☐ in 6 weeks.

We know the total.
We need to find the
number in equal groups.

(b)

$30

?

$30 \div 5 =$

Dion gets $ ☐ a week.

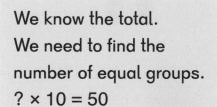

We know the total.
We need to find the
number of equal groups.
$? \times 10 = 50$

(c)

$50

?

$10

$50 \div 10 =$

It will take Emma ☐ weeks to save $50.

Do

1 A store owner put 2 lb of sour balls in each bag.
How many pounds of sour balls are in 7 bags?

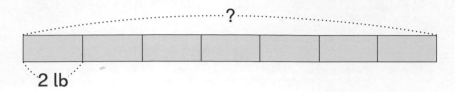

2 Mariam put 14 baseballs equally into 2 bags.
How many baseballs are in each bag?

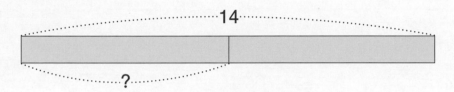

3 Calista bought 8 lb of cherries.
1 lb of cherries costs $5.
How much did she pay?

4 6 boxes have 5 lemons in each box.
Brian wants to put 3 lemons each into bags.

(a) How many lemons does he have?

(b) How many bags will he need?

5 17 children want to divide themselves into 2 teams.
Can they make equal teams?

Exercise 11 • page 177

Review 2

1 (a) 300 + 60 + 9 = ▢

(b) 439 = 400 + ▢ + 9

(c) 500 + 7 = ▢

(d) 895 = ▢ + 95

2 Put the numbers in order from least to greatest.

| 789 | 897 | 978 | 798 | 987 | 879 |

3 Find the value.

(a) 500 + 80

(b) 508 – 304

(c) 800 – 40

(d) 544 + 381

(e) 623 – 218

(f) 487 + 255

(g) 527 – 248

(h) 804 – 317

(i) 700 – 281

4 Find the value.

(a) 7 × 5

(b) 6 × 2

(c) 5 × 4

(d) 2 × 9

(e) 8 × 10

(f) 45 ÷ 5

(g) 16 ÷ 2

(h) 20 ÷ 2

(i) 50 ÷ 10

5 Cooper made 48 medals.
He put 10 medals in each box.

(a) How many boxes of medals does he have with 10 medals each?

(b) How many medals does he have left over?

6 There were 187 students in the programming club.
48 of them left.
How many students are now in the club?

7 A pair of socks costs $2.
How much do 9 pairs of socks cost?

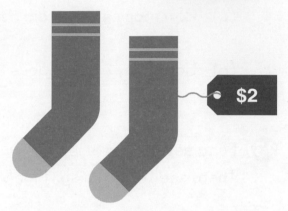

8 There were 300 birds in a tree.
After some flew away, there were
65 birds left in the tree.
How many birds flew away?

9 Austin has 25 cars.
He puts them in 5 rows.
How many cars are in each row?

10 Mei's dog Spot weighs 45 lb.
Her cat weighs 27 lb less than her dog.
What is the total weight
of both pets together?

11 Chapa has 10 nickels.
She spends 4 nickels.

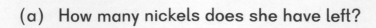

(a) How many nickels does she have left?

(b) How much money does she have left?

12 Gavin bought 3 packs of erasers.
Each pack has 2 erasers.
He gave 3 erasers to his friend.

(a) How many erasers did he have at first?

(b) How many erasers does he have left?

13 Fang saved $5 a week for 9 weeks.
Then, she bought a shirt for $17.

(a) How much money did she save?

(b) How much money did she have left?

14 Aisha had 15 sports cards.
Her sister gave her 20 more sports cards.
Then she put all the cards equally into 5 bags.
How many cards are in each bag?

Exercise 12 • page 181